ETHICS WITHIN
ENGINEERING

ALSO AVAILABLE FROM BLOOMSBURY

The Bloomsbury Companion to Ethics, edited by Christian Miller
The Bloomsbury Companion to the Philosophy of Science,
edited by Steven French and Juha Saatsi
Environmental Ethics, Marion Hourdequin
Philosophy of Science: Key Concepts, Steven French

ETHICS WITHIN ENGINEERING

An Introduction

WADE L. ROBISON

BLOOMSBURY ACADEMIC
LONDON • NEW YORK • OXFORD • NEW DELHI • SYDNEY

BLOOMSBURY ACADEMIC
Bloomsbury Publishing Plc
50 Bedford Square, London, WC1B 3DP, UK
1385 Broadway, New York, NY 10018, USA

BLOOMSBURY, BLOOMSBURY ACADEMIC and the Diana logo are trademarks
of Bloomsbury Publishing Plc

First published in Great Britain 2017
Reprinted 2017, 2020

Cover design: Catherine Wood
Cover image © United States Patent and Trademark Office: http://www.uspto.gov

Bloomsbury Publishing Plc does not have any control over, or responsibility for,
any third-party websites referred to or in this book. All internet addresses given
in this book were correct at the time of going to press. The author and publisher
regret any inconvenience caused if addresses have changed or sites have ceased
to exist, but can accept no responsibility for any such changes.

A catalogue record for this book is available from the British Library.

A catalog record for this book is available from the Library of Congress.

ISBN: HB: 978-1-4742-8604-6
PB: 978-1-4742-8605-3
ePDF: 978-1-4742-8603-9
eBook: 978-1-4742-8606-0

Typeset by Integra Software Services Pvt, Ltd.
Printed and bound in Great Britain

To find out more about our authors and books visit www.bloomsbury.com
and sign up for our newsletters.

To Christina

CONTENTS

LIST OF FIGURES

PREFACE

When I came to the Rochester Institute of Technology as the Ezra A. Hale Professor in Applied Ethics, the Provost suggested I visit the Dean of the College of Business to see if I could help with business ethics. The Dean said, in dismissing me, "We're all ethical here." I was amused, but went next door to the College of Engineering. The Dean there, Paul Petersen, welcomed me, but told me that if I was going to have any street cred among engineers, I needed to take their senior design course, the capstone of the five-year program. So I did, and I learned an immense amount working with a group of students designing and making a self-propelled colonoscope. I learned more about the workings of the colon than I ever wanted to know.

I then started teaching with Jasper Shealy in the Department of Industrial Engineering, an association that continued for four years or so, and afterward lectured on ethics in engineering to the students in the capstone course, telling them something I thought, and think, they needed in their first year.

I cannot thank Paul and Jasper enough for their kindness in letting a philosopher into their midst and to Jasper in particular for letting me teach with him. I found him a wonderful teacher, and I learned far more about engineering than I did about colons. They have my thanks. They would probably think I did not learn enough, but I certainly cannot hold them blameworthy for what follows. They did their best with the material they had, with me, that is.

I also want to thank all my students through the years and especially those to whom I have explained the idea of an error-provocative design. The idea itself seems to provoke example after example, and much of what I have been given by them has found its way into this book. I am particularly indebted to my colleague, David Suits, both for the surfeit of examples he has provided and for his having read through the manuscript and made many a helpful comment.

I am also indebted to Adam Potthast at Park University and Mark Vopat at Youngstown State University. They each used drafts of the book in classes, and I have learned much from their responses and their assessments of where students had problems understanding the text. Just as engineers need to test their design solutions, so writers need to test their creations. Some may decide the book needed more testing, but I alone am responsible for the errors that remain.

My wife, Christina, has been a godsend, helping me talk through problems I ran into as I tried to put my thoughts into words uncluttered by philosophical jargon. I also owe a special thanks to our companions—to my beloved Scout, the wonder pup, now gone after our affectionate fifteen years; to our beloved and much missed Mangia and Gus and Tess, the fierce kitty, who came with Christina; to Raven, our live-in crippled bantam rooster, also now gone, for the companionship he gave us all as well as the insights into just how bright a little rooster can be; and to our new kitty, Peaches, and the pups, Laddie, Gage, and Sunny.

FOREWORD

An interest in engineering ethics has generated an enormous amount of scholarship over the past few decades. So anyone who writes on how ethical considerations enter into engineering owes much to many. But though I have learned much from those who have written on the subject, I will cite few because I will be concentrating on a way in which ethics enters engineering practice that has been downplayed, if not downright ignored, in the vast literature we now have.

I will concentrate on how ethical considerations enter into the intellectual core of engineering, the solution to design problems. Engineering begins with a design problem—how to make occupants of vehicles safer, settling on the interface for operating an X-ray machine, designing more legible road signs. Any design problem leaves much room for creativity and innovation, and so the range of possible solutions to any particular design problem is broad. We can see how broad by looking at the various kinds of cars, or toasters, or coffeemakers, or computers: each artifact marks one design choice over another.

In choosing any particular solution, engineers must make value choices, and, obviously, as we again know from looking at engineering artifacts like cars, not all design choices are equal. Each reflects a particular configuration of values with a particular set of effects, the effects ranging from those produced by obtaining the material from which the artifact is to be manufactured, to those produced in the manufacture, to those produced in moving the artifact to market and storing it until it is sold, to those produced by those who use the artifact, to those produced in disposing of or recycling or remanufacturing the artifact once its useful life is completed.

The easiest way to understand how ethical considerations enter into engineering is to focus on design solutions which cause problems for those who use the artifact embodying the design, and the clearest examples of those are solutions which provoke even the most intelligent, well-trained, and most highly motivated into making mistakes and sometimes causing great harm—by designing an X-ray machine that can easily over-radiate patients or a car or truck with a high risk of exploding if hit.

Everyone is subject to the minimal ethical principle: do no unnecessary harm! Engineers have special obligations to take care not to cause unnecessary harms because they can cause a great deal of harm by virtue of being engineers and are best positioned to choose design solutions that minimize harm.

The intellectual core of engineering, the source of the intellectual joy that animates it, is the working through of various possible design solutions and settling on a particular design that solves the original problem and perhaps pushes the envelope of design. At its core, this is an ethical enterprise since the particular configuration of effects of each design solution will cause more or less harm and so will be, all else being equal, more or less ethical.

These ethical issues are internal to the discipline of engineering. An internal ethical issue is one that arises within a discipline. No one can be an engineer without solving design problems, and so no one can be an engineer without making the ethical decisions we must make in solving those problems. We should presume that every discipline has its internal ethical problems. A physician, for instance, cannot practice without treating patients in one way or another—with respect as a person, or as a piece of machinery to be fixed, say—and those are radically different ethical views to take of a patient.

Such internal problems are distinct from what I call external ethical problems—an engineer who, as a buyer for a company, faces requests from a supplier to let through somewhat questionable parts; an engineer who is upset to find himself working under a younger female boss when he thought he was going to be promoted; an engineer who, as a manager, is ordered by someone farther up the chain to get a product done by a certain time when the testing will not have been completed. These are problems that arise because the engineer is not working just as an engineer, but as a buyer, employee, or manager—positions any professional could hold and problems any professional may face.

Internal ethical issues are those that only someone within a discipline will face, and they are, to my mind, the most important ethical issues engineers will face. Yet, as I say, they tend to be ignored. This book is the antidote to that. I will generally ignore external ethical issues to concentrate upon internal ones.

That is not to say that external ethical issues are unimportant. It is to say that we need to pay at least as much attention to internal ethical issues as the current literature on ethics in engineering tends to pay to external ethical issues.

I came to see the value of thinking of design solutions as embodying ethical choices when I took the Senior Design class at my university. I worked with a number of engineering seniors from various departments in the college, and we had a contract with two internists from the Strong Memorial Hospital in Rochester, New York to develop a self-propelled colonoscope.

The standard way of inspecting for cancer in a person's colon is to insert a stainless steel articulated endoscope with a lens, a hook for grabbing suspect tissue, and a small hose for cleaning off tissue that needs detailed inspection. The endoscope has to traverse the colon and make two sharp turns where the colon attaches to the rib cage on either side, and the risk of harm is high because cancer makes the lining of the colon friable and easily penetrated—especially by a steel endoscope with the circumference of a small pencil. It takes great skill to maneuver the endoscope, and the internists were looking for a device that would significantly decrease the need for a specialist taking extreme care. An endoscope that would propel itself through the colon and do so without touching the colon walls was the goal of our engineering group.

What I noticed was that the engineering students and I were looking at the problem in different ways and so focused on different aspects of our project. The engineering students were intently concerned with getting something that would work. "How do we get it to move through the colon?" I found myself thinking about how the endoscope would be used and so focused on what could go wrong. Since not touching the colon walls was part of the design problem, the students considered that, but failed to consider what would stop this motorized endoscope if it took off up the colon. When that concern was raised, a student said, "Ah, good point," and the group proceeded to ensure that the endoscope could not take off. Their focus put to the periphery of their vision, unnoticed except when drawn to their attention, concerns about the harms to be avoided.

If we focus not just on whether the solution solves the original problem but on whether it solves the problem without causing any unnecessary harms, we make explicit what is implicit in any choice of a design solution: we are making an ethical choice no matter what we choose. Once realized in an artifact, each choice carries with it a set of harms, and except choices with only minor differences, those sets are going to differ from each other. We do not need to provide a formula for weighing those harms against each other or against the benefits that may also be realized to see that, whatever the results, we would be putting on the scales what has moral weight.

Engineers distinguish between what they call the hard and soft, or professional, skills.[1] The former are what students learn from STEM courses, the latter supposedly the "extra" stuff like an ability to communicate effectively. There is a movement afoot to make these skills part of the engineering curriculum.[2] Among these skills, listed in the ABET criteria for all engineering programs, is "an understanding of professional and ethical responsibility."[3]

It turns out that in making use of their hard skills to solve design problems, engineers cannot help but make use of that so-called soft skill. They cannot help but make an ethical choice in choosing one solution over another, I shall argue. So moral considerations are already embedded in the intellectual core of engineering, the solution to design problems.

I shall make this point as vividly as I can by focusing on what I call error-provocative designs to illustrate that ethical considerations enter into design solutions. These are design solutions that provoke errors in even the most intelligent, well-trained, and highly motivated operators in the most pristine circumstances. When an accident occurs, we properly blame the artifact, I shall argue.

Using error-provocative design solutions to illustrate how ethical considerations enter into design solutions may mislead readers into thinking I am writing a book to warn engineers not to pick such terrible design solutions. But I am not looking at what goes wrong, the disasters to be avoided, in order to tell engineers to avoid them—but to illustrate most clearly how any design solution embodies ethical choices and how engineers need to make explicit what they are already doing implicitly in solving any design problem.

The aim of this book, in short, is to show that ethical considerations enter into all design solutions and thus are integral to the intellectual core of engineering. They cannot be avoided. The aim is to make explicit how those ethical considerations enter.

The ultimate goal is to change the way in which ethics is taught in engineering. It is now either an add-on to existing courses, generally discussions of cases, or a separate course called Engineering Ethics. Both alternatives send the message to students and faculty alike that ethical considerations are not integral to engineering practice. I shall argue in what follows that they are.

I obviously do not expect this book to change a long-standing practice, but do hope that once the idea is given a hearing, it will win adherents and ultimately change the practice. That change will require pushing back against the quantification of ABET criteria, but also, in the meantime, providing a numerical weighing, however artificially determined, for the various harms that may occur with various measures of risk for each of them. Engineers are certainly more competent to do that in the detail required for any particular design choice than anyone outside the discipline.

1

Introduction

Morality permeates our lives in the artifacts we use, for they reflect our moral choices. The intellectual core of engineering is the solution to a design problem, and those problems always leave room for creativity. In choosing among possible options, engineers are also necessarily playing off one value against another, and the choice of what to create will necessarily, once realized in an artifact, have effects—some good, some bad. In addition, once the solution is clearly articulated, the artifact that is to realize it follows, and that artifact may incorporate, intentionally or not, new features that cause different effects—some good, some bad. It is a basic moral principle that we should cause no unnecessary harm, and so, in making a choice, an engineer has a moral obligation, at a minimum, to ensure that none of the harms likely to ensue from the artifact are gratuitous.

The most striking examples of how moral considerations enter into design solutions come in error-provocative designs. These are design solutions which are going to provoke errors in even the most intelligent, well-trained, and highly motivated operator in the most pristine of conditions. Since neither the circumstances nor the operator can be at fault, the artifact must be, and we are all familiar with such objects—

doors that appear to open one way but open the other way, for instance. What error-provocative design solutions illustrate is how essential moral judgment is for an engineer engaged in the intellectual core of the profession.

Our moral world

Engineering artifacts permeate our lives—from cars and iPhones to bridges and planes. We are all familiar with things so badly designed that they cause us to make mistakes: doors that look as if they open one way when they open the other way; control knobs that look as if they are to be turned to operate, but must be pushed in or pulled out instead; "DO NOT ENTER" signs on entrance ramps misplaced so they seem to tell us not to enter where we must. It is unfortunately all too easy to find such designs.

We can always find news of them in the headlines. The crash of the Virgin Galactic SpaceShipTwo killed the copilot, who caused the crash when he "prematurely unlocked a section of the space plane's tail used in braking."[4] What is more disturbing is that the company that did the hazard analysis failed to consider "pilot-induced" errors.[5] The company concentrated on the plane and failed to consider how hazards could be introduced through how it would be flown.

The copilot's error was presumably not induced by the plane's design, but it is easy enough to find designs that provoke errors. The worst are those that provoke mistakes for even the most intelligent, well-trained, and highly motivated operator, in the most pristine of circumstances. We can find such designs in even the most mundane artifacts. We need look no farther than our toasters.

One comes packaged with a slip of paper saying, "WARNING! To interrupt toasting, turn toast color control to off/cancel. Do not push the toast lever manually. Internal mechanism will be irreparably damaged." As someone asked, "What kind of toaster is 'irreparably damaged' by using the LEVER to remove the toast?" We use the lever to push the toast down, and levers generally work in both directions: what goes down comes up.

The toaster mechanism will be irreparably damaged by many users who failed to see the warning or, having seen it, pulled the lever up out of habit while hurriedly trying to save the toast from burning.[6]

That toaster is an accident waiting to happen, an unfortunate solution to part of a complex design problem: how can we toast bread and yet interrupt the toasting? Perhaps the solution was driven by considerations of cost or a change in the internals of the toaster, but to the extent that engineers designed the toaster and signed off on the final design, they are responsible for the results—for the predictable harm of customers breaking the toaster, for instance.

A toaster that can be irreparably damaged by lifting the lever up is an artifact whose production was a waste and whose quick end is waste that we must put somewhere. We have in that artifact a set of unnecessary harms—those that come from getting the materials to make it; those that come from squandering the energy required to make it; those required to package it, ship it, store it, and use it until it burns our toast and we break it; and those required to rid ourselves of the trash it has become. These are harms because they set back interests we have in, for example, not wasting our money on something that will quickly break and, for another example, in not polluting our air and groundwater any more than necessary. For engineers to choose that particular toaster design from all the possible designs is to make a moral choice, one that will produce more harms, and worse harms, when realized in an artifact than other choices that could have been made. We live in a contingent world that reflects moral choices we have made.

We each no doubt have our own favorite examples. They seem to be object lessons in the frustrations of life, things we have to live with. But there is no necessity that toasters be designed that way or that "DO NOT ENTER" signs be so placed as to mislead drivers into thinking they are on the wrong ramp or that doors look as though they open one way when they open the other. These examples come about because of the choices people made. They are artifacts, designed and created by us.

If it seems puzzling that ethical considerations enter our lives even in the artifacts with which we have populated them, think of how ethical considerations enter our lives even in what we might consider the most mundane of circumstances because of choices we make. If we choose to pick up and answer our cell phone while driving, we have chosen to increase the risk of our having an accident as well as the risk to others. Increasing the risk of harm is itself a harm, and so, in choosing to answer the cell phone, we have chosen an option that is more harmful than the other option, immediately available to us, of not answering the phone.

We have few better examples of how our lives are shaped by such decisions. Few, if any, who drive have escaped having to shape their driving by another driver's failure to signal because preoccupied with a cell phone and without a free hand or by the slowing down and speeding up as the driver gets more or less animated while talking. The list is long, but the point is short: the way we move down the highway is no different from the way we move through the world. We move in a world created and shaped by moral decisions.

So we should not be puzzled that morality permeates our lives through the artifacts of our lives—from the toaster we face in the morning to the ramp on the thruway with the "DO NOT ENTER" sign. These artifacts are the result of choices made by those who designed them.

The path from a design problem to its solution to its realization in an artifact is a complex one, with many a possibility of error along the way and many a way for the initial design solution to be altered by someone other than the designer in order to save costs, for instance. I will concentrate not on the path, but on the initial step, the solution to a design problem.

We find in such a solution the intellectual core of engineering, and although it is often claimed that engineering is a purely quantitative discipline, it is not. Ethical considerations are at the core of engineering. They are essential to engineering practice. Remove them, and we cannot have engineering.

Design problems

A condition of our doing something moral is that we could have done otherwise.

When toddlers trip and fall, we comfort them; when they throw themselves on the ground in a temper tantrum, we at the least look askance. Engineers can make a moral choice in picking a design solution because there is no single way to solve any design problem. A statement of a design problem, however detailed, does not necessitate any one solution.

We need only consider toasters and the myriad forms they can take or, for that matter, toothpicks.

The initial statement can be sparse: design a pick to get food and other such things out of your teeth. "Ah, a toothpick! What could be easier?" We may well wonder how there can be much room for creativity with such a design problem. How many possible different kinds of toothpicks can there be? And how could any value choices influence the answer, especially moral values?

We can see an answer to that question in the toothpick given in Figure 1.1:

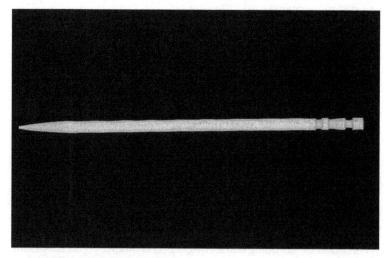

FIGURE 1.1 *Japanese toothpick.*

This is a Japanese variation of a toothpick, pointed at one end with "a series of grooves encircling the toothpick" at the other end.

Once you use the toothpick, you are to break off the end at one of the grooves. You then place the end, like a Japanese pillow, on the table, with the rest of the toothpick resting on it, pointed end up. That way others can see that the toothpick has been used—a health benefit—and with the used end up "what had been in the diner's mouth does not touch the common table." The design is thus not just decorative, as we might assume, but a clever solution to ensure that a used toothpick is no risk to anyone's health.[7]

This Japanese variation solves a problem not in the design problem with which we began this section: what are we to do with the toothpick after it is used so that others will not use it? An easy way to transfer disease from person to person is to use a common toothpick. So ensuring that a toothpick is used but once is of some importance.

The design does waste wood, however. It will take two of them to provide two pointed ends for picking. But the value of not spreading disease was judged of more value than making full use of a piece of wood for picking. That is arguably a moral judgment since the aim is to mitigate the harms that come from spreading germs through using someone else's toothpick. The design thus expresses a set of values.

The design is also an example of another feature of design problems. As it turns out, initial statements of design problems inevitably go through a transformation as engineers work out what might and might not work, what causes additional problems (such as puncturing your gums or damaging the enamel on your teeth), what can be readily manufactured, what can be manufactured cheaply enough to make it commercially viable, what problems are missed by a design, and so on.

In *The Toothpick: Technology and Culture*, Henry Petroski details a variety of transformations of the design problem. That initial sparse description for a toothpick ends up including something like this:

> These areas between adjacent contacting teeth, i.e., the interdental spaces and the interproximal tunnels, are actually like a passageway with a somewhat triangular cross-sectional shape. The base of the triangle is the gum or gingival tissue; the sides of the triangle are the proximal surfaces or side walls of the contacting teeth; and the apex of the triangle is the incisal or occusal contact area of the two adjacent teeth.
>
> Quite often the openings to these tunnels and spaces are blocked by slightly swollen or edematous gum tissue. Therefore, in order to enter the spaces or tunnels, the cleaning instrument must be sufficiently resistant to bending perpendicular to its longitudinal axis to enable it to depress or displace the gum tissue blocking the entrance or exit to the tunnels or spaces. Furthermore, the posterior interproximal tunnels are often quite tortuous, i.e., the path of the passageway is circuitous.
>
> Therefore, the instrument must be sufficiently bendable to follow this tortuous tunnel as it contacts the hard surfaces of the teeth and firm healthy gingival tissues. It must also have sufficient strength to dislodge food debris and loosely adherent calcular material from the walls of the tunnel or space. It must also intimately conform to the walls of the sides of the tunnels and spaces and must have sufficient abrasiveness to remove the dental plaque without injuring the tooth or gum tissues. Additionally, it must be able to fit into the usually narrow space between the anterior teeth.[8]

Who would have thought that designing a simple toothpick would require such a detailing of the work a pick would have to do? And this description does not even cover concerns about ease of manufacture, the availability of material, the cost of production, and other such matters that an engineer needs to consider before settling on a particular design solution.

Yet, however detailed, nothing in this design statement determines any particular solution. Even a more extended statement is not going to determine a conclusion. We are not working with a mathematical problem here where the premises determine the conclusion as in, to use the simplest of examples, 2 + 2 determines the conclusion, 4. Any solution will be constrained by quantitative considerations, of course. Not any object can serve as a toothpick. A dandelion stalk is straight, but too flimsy to do any picking; a titanium shaft dusted with industrial diamonds will certainly do a lot of picking, but endanger our gums and enamel. Presumably we could quantify the range of stiffness permitted, a range that would exclude the dandelion shoot as not stiff enough and the titanium shaft as too stiff. An engineer has to take such matters into consideration, but one feature of such a design statement as that for a toothpick is how much conceptual space it leaves open for solutions. Even the simplest of objects, that is, can have many different variations, and that means that no design statement determines its solution.

The intellectual core of engineering, and the source of the joy of success, is the working through of various possibilities and settling on a particular design that both solves the original problem and, where possible, pushes the envelope of design (and is, perhaps, aesthetically pleasing as well).

So we end up with flat toothpicks and round ones, with toothpicks pointed at one end to toothpicks pointed at both, and even a toothpick that fits on the end of one's tongue as in Figure 1.2.[9]

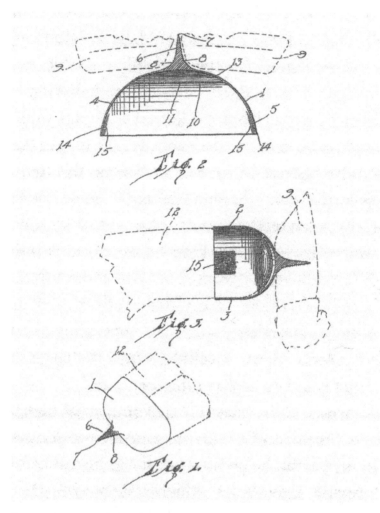

FIGURE 1.2 *Toothpick for the tongue.*

Who would have thought? Human ingenuity knows few bounds.

Design problems are subject to extension and modification, that is, as various possible solutions are considered, their strengths and weaknesses assessed, and new possible features are considered and incorporated into the original design problem. Our inability to reach certain "interdental spaces and the interproximal tunnels" easily was presumably a consideration for the odd tongue toothpick. Its inability to reach the front of our teeth readily, and a serious concern about accidentally swallowing it as we probe and pick and push, would certainly be considerations in deciding whether to use it in place of more familiar solutions to the design problem.

Conceptual space for creative solutions

What is most important for our concerns here is that though the design problem constrains potential solutions, it leaves open enormous space for creativity. Engineers are in no different position than, say, poets in this regard. A poet is constrained by prior choices, both the poet's and those of others. We can no longer say, "It's a slam dunk!," without invoking for many George Tenet's mistaken response to the question whether Iraq had nuclear weapons. The phrase resonates differently now than it did before Tenet uttered it in that context. That new resonance is as much a constraint on a poet's choice of words as are, for instance, the meter chosen, the rhymes and rhythms of various words used, and the subject matter, and the point, of the poem. These are not quantitative constraints, of course—though the meter may be—but they constrain the creative genius of a poet just as much as, and no more than, a design statement and quantitative considerations constrain an engineer's choice of a design solution.

In both cases conceptual space exists for creative solutions, and engineers who think themselves immune from considerations of value because they are in a realm of crystalline quantitative clarity misdescribe the intellectual core of their discipline. It is as though they are taking the quantitative constraints of a design problem not as constraints on the problem, but as the only matter of concern to engineers. But the intellectual core of engineering—the intellectually exciting part of the discipline—is the solution to a design problem, and those solutions are not determined wholly by quantitative considerations. Engineering design solutions do not bear the same relation to an engineering design problem that mathematical conclusions bear to their premises. A creative mathematician may find an elegant way of deducing the proper conclusion from the premises, but however creative a mathematician may be, the chain of inference is deductive, and the statement of the problem leaves no conceptual space for a different conclusion: 2 + 2 equals 4 no matter how one arrives at that conclusion from those premises.

The relation between an engineering design problem and a solution is mediated, however, not by deductive inferences, but by a creative mind capable of imagining different ways of solving the problem and equally capable of choosing between those different solutions, weighing the advantages and disadvantages of each possible solution, and making a wise choice. As Petroski, among others, has said, a design problem does not determine its solution.[10] Fastening on a particular design solution is a two-stage process, that is. In the first stage, we brainstorm alternative solutions to the problem, and in the second stage, we choose one of the alternatives, preferably culling out those features which are not essential to the design's resolving the problem.

As the toothpick examples illustrate, a design choice reflects value considerations.

The Japanese toothpick design ranks the healthy disposal of a toothpick above the convenience of having two pointed ends with which to work. The toothpick that fits on the end of a tongue values the capacity to reach into odd corners inside the teeth higher than, say, the oddity of having such an appendage on one's tongue, the difficulty of trying to keep it there, and the risk of swallowing it by accident.

Different choices not only play off different values against one another, but also produce different effects when realized in an artifact. Someone trying to use a toothpick with only one useful end may need more toothpicks than someone with two ends to use. That will mean more toothpicks manufactured with more trees cut down, more waste disposed of, more toothpicks packaged and carried to market, more toothpicks purchased—all effects of a simple design choice: one useful end or two?

So not all design choices are equal, obviously. Each reflects a particular configuration of values with a particular set of effects, the effects ranging from those produced by obtaining the material from which the artifact is to be manufactured, to those produced in the manufacture, to those produced in moving the artifact to market and storing it until it is sold, to those produced in disposing of or recycling or remanufacturing the artifact once its useful life is completed. Not all artifacts are susceptible to all these effects, obviously.

We do not remanufacture toothpicks, for instance. But laying out the possible range of effects allows us to see that in picking any one design solution, we are picking out not only one array of values over another, and one set of effects over another, but one set of harms over another.

As we can see from the tongue-mounted toothpick, some design solutions are more likely to cause or risk harm than others, and in some cases the harm can be more than a trivial annoyance. We should rank a tongue-mounted toothpick fairly far down the list of viable solutions. After all, swallowing a pointed implement large enough to fit on the end of your tongue and sharp enough to pick your teeth is not a trivial matter. It would be a matter of even more concern if the engineer failed to craft the details of the tongue-mounted toothpick so that it would fit tightly on a tongue and not slip off easily, and that problem is not simple either since there are, no doubt, differently sized tongues, longer and shorter, thicker and thinner, requiring smaller and larger toothpicks of varying widths. There may also be differently shaped tongues, some unable to hold onto, as it were, the variant pictured in the patent application. So choosing the tongue-mounted toothpick as a design solution is to choose a design with many possible unnecessary harms. We are lucky other design solutions are possible.

Ethics in engineering

As soon as possible harm enters, ethics enters. Engineers are in a position to do great harm—by designing a bridge that will soon fail, an X-ray machine that will overradiate patients, a car or truck with a high risk of exploding if hit. Everyone is subject to the minimal ethical principle: do no unnecessary harm! Engineers are not immune from that principle. Indeed, they have a special obligation to take care not to cause unnecessary harm because they are in a position, by virtue of being engineers, to cause—and mitigate—a great deal of harm.

When engineers solve design problems, they necessarily invoke that minimal ethical principle. They do so both because their choice plays off values one against another, some of which are moral, and because their different choices produce different configurations of harms and benefits once their choice is realized in an artifact.

But there is a third way in which harms can enter via an engineer's design solution.

Once engineers have chosen a design solution, they may fail to execute it well so it can make its way through the transformation from idea to artifact. Engineers are no different from, say, artists in this regard. There is the conception and then the execution.

A great artist has great ideas executed in a stunning way. These two features of an artist—the capacity for creative ideas and the ability to execute those ideas—are distinct. One can exist without the other. Someone may have wonderful ideas, but lousy technique. Someone may have great technique, but lousy ideas. Neither can be a great artist. A poorly written book with a great plot is still poorly written, and a well-written book with a poor plot still has a poor plot. Just so, an engineer may be able to toss off great design solutions, but fail at doing the detailed work needed to turn the idea into a viable solution or may be superb at the detail work, but fail to think of good design solutions.

These two features of any design problem—a conception of a solution and the proper execution of that solution—map onto two kinds of ethical relations someone takes on in becoming an engineer.

1. **Role morality**—In becoming an engineer, a person takes on a set of role-specific relations having to do with the practice of engineering—e.g., ensuring that calculations are made correctly.

The role carries with it moral obligations. It is no small matter that calculations must be made correctly. Engineers are not allowed to guess how thick a road surface must be to withstand the expected traffic. They are required, as engineers, to calculate what is required. That is a moral obligation because of the harms that can occur if it is not done correctly—accidents and their accompanying harms from a road's breaking up, for instance, and the expense of tearing up and then rebuilding the road surface to the proper specifications.

Professors teach engineering students that they must double-check their calculations. A careless engineer can cause structural failure, with loss of life and other harms. The engineer's intent is as ethically irrelevant as the intent of any professional acting professionally: the daydreaming dentist who drills through one of my teeth is acting unprofessionally, and unethically, despite having no intent to cause me harm. The role-specific ethical relations of the engineer come with the territory independently of whether the student wishes to take them on or is even aware of them.

Such relations are neither trivial nor unusual. They define a profession's ethical core: physicians cure patients, accountants analyze financial information, and architects design buildings. In being taught the profession, students are being taught its core ethical values. When they become professionals, they enter into the set of moral relations that define the profession, and when they practice their professions, they are to act morally in realizing their roles in those moral relations.

Many of the ethical principles central to being an engineer are thus common to other professional disciplines. An engineer should take due care making calculations, but so should my accountant working on my tax returns. The differences between professional disciplines lie in their intellectual cores: what does someone do as a dentist, as an accountant, or as an engineer?

The intellectual core of engineering is the solution of design problems of a certain sort. Figuring out how to make occupants of vehicles safer when accidents occur, determining how to make a click on a keyboard become a symbol on a screen, deciding how to design legible road signs, ascertaining how best to build a bridge, settling on the interface for operating an X-ray machine—the range of engineering design problems is vast.

To repeat what should now be obvious, the range of possible solutions for any particular design problem is vast as well, and in choosing among the possible solutions, engineers must make value choices. Sometimes cost is of more value than aesthetic appeal; sometimes effectiveness wins over cost; and sometimes reliability wins over ease of manufacture. In making these value choices, an engineer is necessarily making moral choices.

Whatever design solution an engineer proposes to an engineering problem, its realization in an artifact will have its effects in the world, producing more or less good, causing more or less harm— through obtaining the material chosen to make the artifact, through manufacturing the artifact, through moving the artifact from where it is manufactured to where it is to be sold, through the use of the artifact after it is sold, through the disposal of the artifact after its useful life is over.

Assessing alternative solutions by weighing the harms and benefits they would produce is complicated if only because we can never be sure what the effects of a solution will be once realized in an artifact.

We shall focus on one aspect of that complicated assessment, the potential harms, because that allows us to concentrate on a moral aspect of engineering that all engineers ought to agree on: between two design solutions, the one that produces less harm is morally preferable, everything else being equal. Other values may end up trumping the morally preferable design solution, but that does not change the moral agreement that engineers ought to avoid, if they can, those design solutions that produce more harm than their alternatives. We should always ask, that is, "Could a different design solution have produced less harm?"

But that question presupposes that we have moved from a design solution to its realization in an artifact, and that movement requires that an engineer properly execute the details of the solution. A mistaken calculation, a misjudgment about the kind of material to be used, a failure to see how a change at one stage reverberates through the rest of the design—all these and more are problems that can occur because an engineer has failed to follow through on a design solution to make its realization possible.

We can expect failures. Engineers are as human as the rest of us, after all, but if they fail consistently, or spectacularly in what should be a simple engineering matter, we judge them incompetent—or not an engineer at all. Engineers must learn a particular set of rules of skill and cannot be engineers without learning those rules. They are then to use those rules of skill with competence and care.

An engineer who fails to execute design solutions with the accuracy in detail needed has failed to fulfill the role-morality of an engineer. Even with the best of ideas about how to solve a problem, that is, an engineer may fail to work it out properly, producing, for instance, a tongue-mounted toothpick that will not stay mounted, increasing the risk of a person's swallowing it by accident. A wooden toothpick must be thick enough and supple enough to withstand the picking and probing of teeth, but thin enough to fit between teeth and strong enough to pick out food without bending so much that the point cannot keep a purchase. It must be long enough to allow the point to reach our back teeth, but short enough that we can use it without difficulty. Length and width are quantitative features of a wooden toothpick, and an engineer needs to get them right. A failure is likely to introduce harms that could otherwise be avoided.

Consider that Japanese variation of the toothpick, the one with grooves around one end. There is the idea of such a toothpick and then the toothpick itself, the artifact that realizes that idea. Between the two an engineer must lay out, put to paper as it were, the original idea and make all the decisions needed so that the toothpick can be manufactured. Those decisions will include such matters as, say, how deep to make the grooves so that the end can be readily broken off when the toothpick has been used and yet not break off when being used. An engineer may fail to do that well, and that failure is an engineering failure, a failure of an engineer to fulfill the role the engineer has taken on in becoming an engineer. It is the equivalent of a physician making the proper diagnosis of a patient's illness and then not figuring out how to treat it properly.

We run afoul of engineering artifacts, a door that does not open the way it appears it ought to open, for instance, and any of

a number of things may have gone wrong in the process of creating
that artifact:

a. The design problem may not have been articulated well.

b. Brainstorming possible solutions may not have produced a
good pool of solutions.

c. The solution chosen from a pool may not have been the best.

d. The detail work necessary to make it clear how the solution
would work to solve the problem may not have been done well.

e. The engineering artifact may fail to realize properly the
detailed solution.

We shall be looking at examples where it is not clear which of these
problems is responsible for the engineering artifact that is causing
problems. Was the design solution problematic? Was it problematic
because the design problem was not properly understood? Did the
engineer make a bad choice from the pool of solutions? Did the
engineer not do the detail work properly so that errors introduced
there continued on their way to the final engineering artifact? Or did
some slip occur between the design solution's execution on paper and
its realization in an artifact?

If we are to correct the problem, it will be a matter of no small
importance to determine exactly where things went wrong. However,
our main concern is not to fix blame, but to show that something
can go wrong at any point in the design process and that engineers
could produce problematic designs through a failure of imagination
in brainstorming ideas, a failure to make a wise choice among those
ideas, or a failure to work out well the details of the solution chosen.

It is (d) that concerns us here, putting an idea to paper, and it is a failure at that point in the process that carries a concern about the role-morality of an engineer. Nothing in an engineer's role can guarantee that a brilliant idea brilliantly put to paper will be realized in an artifact in the way envisaged, but it is the engineer's job, part of the role of engineers, to get the details right—ensure that the measurements are correct, choose the proper material for the artifact, make sure that the parts work together, and so on.

So we have now noted three ways in which ethical considerations enter into engineering—through the execution of a design solution chosen from among all the possible solutions and thus reflecting one set of values over others and realizing one set of effects over others when manifested in the resulting artifact. The execution of a design solution invokes one aspect of an engineer's role-morality, in particular, the quantitative skills that engineers need to learn, but among the moral relations an engineer takes on in becoming an engineer, one set—design solutions—has a special status, and deserves separate consideration, because it defines the intellectual core of the engineer, capturing engineering's creative center.

2. **Design solutions**—The intellectual core of engineering is solving design problems, and because at a minimum, ethically, an engineer ought to cause no unnecessary harm, solving design problems requires ethical considerations if only to avoid a solution which causes unnecessary harm.

An engineer cannot avoid the role-specific ethical relations of being an engineer and the ethical relations connected with solving design problems. A person cannot be an engineer without acting on these ethical relations. They are internal to the profession. The best way to make that point is to consider what I call error-provocative designs.

Error-provocative designs

We know that sometimes the fault is ours when things do not work the way we expect. We push the wrong button on the TV remote because we are not paying attention. Sometimes the fault is in the artifact. The remote's buttons are so close together that we have difficulty pushing one button without pushing another. Or, circumstances being what they are with a small child around, a drink has spilled on it, and the buttons stick. As we all know, technological artifacts—television remotes and cell phones—are among modern life's most ubiquitous little annoyances.

We think of intelligence and training and motivation as protection against mistakes. There is no designing any artifact to prevent a person from making a mistake when the person lacks the intellectual ability to learn how to operate it, but surely, we think, the more intelligent someone is, the less likely the person is to be misled by some feature of the artifact. There is no designing any artifact to prevent those who are untrained from making a mistake, but surely the more highly trained someone is, we think, the less likely the person is to be misled. There is no designing any artifact to prevent those who are inattentive or distracted from making a mistake, but if the person is intelligent, well trained, and paying careful attention, we think, the less likely the person is to make a mistake.

What makes error-provocative designs unique is that high intelligence, the best training, and the highest of motivations make no difference: the artifact is going to provoke the person into making a mistake. It does not just permit those who use it to make mistakes or even encourage them to make mistakes. Something about the design provokes them into making mistakes. We can even imagine that the worse of error-provocative designs makes use of a person's intelligence, training, and motivation to provoke errors. The more intelligent and more highly trained and motivated an individual is, the more likely that individual is to make a mistake. The design would make thinking about it and past training and even high motivation impediments to using it without error.

We need only think of a design solution that looks as though it works the way all previous artifacts like it have worked, but in fact will break if someone intelligent enough to know what to do and well trained in using such artifacts is highly motivated to use it. The toaster with the lever that will irreparably break the mechanism if pushed or pulled up is an example of such a design solution. Dumb luck would be an asset in using such artifacts.

It is sometimes said that once we get a word or phrase for something, we can see instances all around us. This is true of the concept of an error-provocative design. It is easy to find examples. For instance, in the public library in the town in which I live, the knob on the door to the men's room on the second floor is on the right-hand side of the door, but opens the door only if turned left, counterclockwise. That is such an unexpected way for a knob to turn that the door at one time had a handwritten sign on it that said "Knob turns to the *LEFT*!!!" I assume the sign was put there by a librarian who was tired of telling men, "No, the bathroom is *not* locked. Just turn the knob the other way!" A knob that must be turned in an unexpected way to open the door is a real novelty and sure to provoke errors.

The men who pestered the librarian were all presumably relatively intelligent. They were in the library after all. We can also presume that they were well trained in opening doors. All of us have been opening doors, or trying to, at least since we first began to walk. And they were motivated. It is a bathroom door. It is a measure of how unusual it is to have a door knob that opens a door by being turned in an unexpected way that so many men went to the librarian complaining that the door was locked.

You can put yourself in their place by imagining a door handle rather than a knob. Suppose that the door handle is on the right-hand side of the door. Put your left hand out to open the imaginary door by grabbing the handle and turning. What did you do? You reached your hand out, with your knuckles on top, and turned the handle down, counterclockwise, ready to pull it open so you can then walk through. But suppose the door opens only by being turned upward, clockwise. Try making that motion with your hand—still with your left hand reaching across in front of you and in the same position, knuckles up. It is difficult to open a door that way even if you knew that was how it opens, and when you expect it to open by a downward motion, as door handles normally do, it would not likely occur to you to try to open it by pulling the handle up. It is not only difficult to open a door that way. If you did open it that way, your hand would pull your arm and shoulder out of position to get through the open door without grabbing it with your other hand. That is one reason you would assume the door was locked. Having a door handle open that way is an extremely poor choice given the usual way we open and go through doors.

Putting such handles on fire exits would be catastrophic in an emergency. People would run to open the doors and think them locked when they were unable to turn the handles down. Such a handle would be as catastrophic as having an emergency door that opened in rather than out. Both are error-provocative and will produce catastrophic results in such a context.

It is morally wrong to put individuals at an unnecessary risk of harm, and so any company that installed handles in that manner, and any person who supervised the installation, would be morally culpable. The problem would probably lie in the installation, not in the design of the handle itself. So we would not fault the designer unless something about the design itself made installing it in the wrong way necessary or likely.

But we know there is plenty of room in every design problem for a solution which provokes errors on the part of even the most intelligent, well-trained, and highly motivated operator. We will focus on error-provocative design solutions because they illustrate most clearly how ethical considerations must enter into the intellectual core of engineering—if only to avoid such design solutions.

That artifacts can be so badly designed as to cause moral problems runs counter to a standard view that ethical considerations are extraneous to engineering practice. On this standard view, ethical issues arise about what people do with what engineers create, but engineers are not responsible for the use others make of what they do. The engineers who design a car are not responsible if a driver runs down a pedestrian any more than the engineers who design a pen are responsible for a child poking out an eye. But if even the brightest, most well-trained, and most attentive are provoked into making mistakes because of an engineer's design solution, the engineer can hardly fault the user.

We could imagine an engineer intentionally producing such a design solution—to get back at an employer for perceived mistreatment perhaps. Were such a design solution realized in an artifact, we would produce a form of entrapment: the user would be led to make a mistake that the design itself provokes. We would hardly blame the user for having been misled into a mistake. We would blame the engineer for having set the user up. The concept of an error-provocative design is meant to show that the standard view of engineering is mistaken.

On that standard view, engineering rests in a purely quantitative world, completely separated from the messy world of our lives where we use what engineers create to do good and evil. Engineering has nothing to do with that, the standard view goes.

Some engineers might concede that what they create can have ethical implications regarding, for instance, pollution, but think that engineers should concentrate on those aspects of their practice which permit quantitative answers and let others consider whatever may be the qualitative and so contentious moral implications of their work.

However reasonable these responses may seem, they rest upon a mistaken contrast between engineering and ethics and also upon mistaken understandings of both engineering and ethics. An engineer's decision about what to do to solve a particular design problem does not rest wholly on the crystalline clarity that quantification provides, but on value considerations and ethical considerations as well. We cannot deduce from anything in a design problem any one particular solution. Thinking up possible solutions and picking one over another calls for something other than deduction, that is, an imaginative response. An engineer will need to create and choose a solution, not just look up the solution after doing all the relevant calculations.

Engineers may well blanche at the idea. For, they may think, if the design solution depends, even in a small part, on ethical considerations, something qualitative, vague, subjective, and contentious will have found its way into that pristine quantitative realm they think is the heart of engineering.

That concern is understandable, but misplaced, and particularly so if the worry is that I am suggesting ethics ought to be introduced into engineering. I am not arguing that ethics ought to enter into the heart of engineering. I am pointing out that it is already integral to the solution of the design problems that form the intellectual heart of engineering. I am only describing what they already do. That new description does not change what they do or distort it in any way, but brings to light a feature of what they do that has been bleached out by a mistaken understanding of what they do.

In short, crucial judgments engineers make in solving design problems are ethical judgments, whether the engineers realize it or not. This is not to suggest that engineers are unethical. Quite the contrary. That we have so few accidents in our complex technological world shows that they are ethical. Still, some design judgments are mistaken, and so accidents occur.

We shall consider in the next chapter how we analyze accidents. The aim is to sort out the contribution of the artifact involved: was there something about its design that contributed to the accident?

We shall then examine an accident in Chapter 3 and see how a faulty design was causally responsible for the loss of 159 lives. Nothing about the design problem necessitated the engineers solving it as they did, and the way they solved it was bound to produce an accident. Their design solution was error-provocative and did indeed provoke an error, causing unnecessary harm.

2

Analyzing accidents

Whenever there is an accident, among the first things we hear is that it was an operator error. Either the operator did not have the intellectual ability to learn how to operate the machine (the airplane, the train, the automobile), was not well-trained, or was not paying attention. But there are two other variables in any accident besides the operator. The circumstances in which the accident occurred may be causally relevant. Even the best of drivers, paying full attention, can have an accident on black ice, but when the circumstances are perfect and the operator in the zone, as it were, we have to lay the fault on the artifact—something wrong with the machine that caused even the best to make an error, resulting in an accident. Such artifacts are the result of error-provocative designs, and although we should obviously avoid such design solutions, the lesson they teach us is that moral considerations enter into every design solution because any solution may produce avoidable harms.

What can go wrong

Accidents tell us how to do things better—provided, of course, that we find out what went wrong. When we have an auto accident, for instance, the problem may lie with something the driver did or neglected to do, with some unusual feature of the situation, with the artifact in question, or with some combination of these three variables. These variables—the operator, the circumstances, the artifact in question—must be examined in any accident. An analysis will become complicated when more variables are involved—a copilot as well as a pilot, for instance, as in the Virgin Galactic SpaceShipTwo disaster. We will concentrate on the simplest kind of accident where a driver may have hit another car because the brakes failed, or the driver was distracted and failed to stop, or black ice made the brakes useless. Whether it is the artifact, the operator, or the circumstances will make all the difference in trying to prevent a repetition.

1. **Operator**—We cry "Operator error!" if the operator

    - does not have the intellectual ability to learn what needs to be done,

    - has the intellectual ability, but was not well trained, or

    - was well trained, but off in some way.

Each of these three possibilities covers a wide variety of kinds of failure. We can fail to be off in some way, for instance, for many different kinds of reasons.

Think of people driving. They may fail to avoid another car because they are distracted (by a bird just missing a windshield, for instance), engaged in something else that requires too much of their attention (talking on a cell phone, turning to chastise a child in the back seat, fiddling with a phone to text message), drunk or high and so unable to concentrate fully on what they ought to be doing, angry and so thinking about something completely different from what they should be thinking about, and so on. They may be preoccupied—as were the NWA pilots who flew 500 miles without radio contact and overflew their landing site in Minneapolis by 150 miles because, they said, they were working on their computers.[11] They may be asleep—as perhaps were the pilots on a Go! Airlines flight that at 21,000 feet was out of radio contact for 25 minutes, flying 15 miles out to sea past Hilo, its landing site, before turning around and landing.[12]

The phrase "off in some way" is meant to cover all the variety of ways in which we can fail to engage fully in what we are doing even though we have the intellectual ability to have learned what needs to be done and have been well trained. Even the most intelligent and well-trained people can still be distracted or find their minds wandering or, as it were, inoperative at crucial times. In investigating an accident, we must come to grips with all these possibilities, a difficult matter in any event, but especially if the operator has died or was plagued with more than one problem.

We are all familiar with being on top of our game, as it were, in the zone where we can do no wrong. Either we have experienced it or read about or saw someone for whom everything went right. We may wish that experience were not so rare as we sometimes stumble our way through life, but the point of the phrase "off in some way" is to capture all the ways in which we can stumble. There seems to be no general term available to cover all the ways in which we can fail.

"Unmotivated," "distracted," "out of it," "inattentive," "absentminded,"
"drugged," "sleepy"—the list is long and obviously covers a great many
different ways in which we can fail to be fully engaged with what we
are doing. I mean to cover all those possibilities with that phrase "off
in some way."

Poor training? Training can go wrong for any of a number of
reasons. It is difficult to train us out of a habit. Especially in times
of stress, the habitual reaction is likely to take over. It is easier to
train us to follow simple instructions than complicated ones. This
is particularly true for instructions that tell us to do one thing most
of the time but something else in one particular circumstance. Even
when instructions seem easy, they may be misunderstood. Everything
that can go wrong with communication can go wrong with training,
and that covers such a variety of failures it is not possible to guard
ourselves against them all. A set of instructions always has a design,
and the design itself can be better or worse, helping or hindering
understanding.

So in investigating an accident, we must examine exactly how
the operator was trained. When an accident is in the offing, was the
training good enough that the operator will know what needs to be
done? In investigating the crash of TransAsia Flight 235, investigators
heard the captain say on the flight recorder when one of its two
engines flamed out, "Wow, pulled back the wrong side throttle"—
an odd comment given that he thereby condemned the plane to a
crash that would kill him and 43 others. The plane was designed
to fly on one engine, but the captain killed the working engine. It
turns out that the pilot had initially failed that part of the training
where pilots have to respond to an engine loss, showing "insufficient

knowledge leading to hesitations in 'both EEC (electronic engine controls) failure' and 'engine failure after V1-situation" where "Vi is the speed beyond which takeoff can no longer be safely aborted." He later passed, but we will never know whether he was not sufficiently trained or, although trained well enough, was off in some way. We have here a good example of how difficult it can sometimes be to sort out exactly what goes wrong, but nothing was wrong with the plane that a well-trained and attentive pilot could not have handled without crashing it.[13]

The aim in assessing the quality of training is to determine if the training needs to be modified in any way. Those who pass through training ideally ought not to fail what they were trained to do when the expertise they supposedly gained is needed. In some cases, obviously, the training itself can contribute to the accident and so needs to be corrected. We might even find that the training was counterproductive.

Buddy Holly died in 1959 in a plane crash that killed all aboard. The pilot was relying on instruments because visibility was limited. He had had "a little bit of instrument training" and so was "not totally unprepared," according to Bruce Landsberg, executive director of the Aircraft Owners and Pilots Association. But, he says, "The instrument that was installed on the aircraft read differently than the instrument he had trained on. So if the aircraft was making a right turn, it would appear on this instrument to be making a left turn—which makes it very difficult to sort things out quickly when you're close to the ground and in moderate turbulence." The crash occurred because the pilot was "not able to keep the airplane upright by reference to the flight instruments." Here the training was counterproductive, teaching the pilot to do exactly the opposite of what he would need to do to keep the plane upright.[14]

Not intellectually capable of understanding what needs to be done? We know that even the brightest people can get things wrong—locking ourselves out of our car, for instance. Scratch a genius, and you will undoubtedly get a story of a silly blunder. So even with a brilliant operator, doing something wrong may have been a factor in an accident. Just so, we know that even those not so gifted intellectually may have excellent common sense and moments of genius. So when investigating an accident, we cannot draw any specific inference simply from a person's general intellectual level. We will need to look out for aberrant behavior.

One difficulty in investigating an accident is that behavior can be so outside the bounds of what we would consider normal that the possibility will not have occurred to us. "How could anyone make *that* kind of mistake?!" This is a source of delight—and horror—at the Darwin Awards: how could anyone have thought to play Russian Roulette with an automatic?[15]

These three possibilities—intelligence, training, and being on— work in tandem. If we determine that an operator has the intellectual ability to learn what needs to be done and is well trained, we focus on the operator's condition to see if something about that was causally relevant. Distracted? Tired? Depressed? If we determine that the operator is intelligent and is fully engaged, we hone in on the training. Did the operator have enough hours using that machine? Can we be confident the training was sufficient? Was the training course thorough? Was the operator hurried through it or given time to understand fully the various kinds of problems that can arise using that machine? If we determine that the operator was well trained and fully engaged, we hone in on the operator's intellectual ability. With all the will in the world, people without the intellectual ability to learn what needs to be done are not likely to make the most of their training and are more likely to make mistakes. Is that what we have here?

Making a judgment about why we have done something is never easy. Even the most commonplace of actions may have multiple motivations. We eat that particular dinner because we enjoy it, because we are hungry, because it is healthy, because someone made it for us and we cannot well refuse to eat without being impolite, and so on. Determining which motivation or how many motivations were causally significant can be difficult for the person eating the dinner, let alone for someone else observing the behavior. It is harder still when we must make a judgment about whether an operator in an accident is somehow responsible and, if so, how. We have to look not only at the many different possible motivations but at the person's general intelligence and preparedness and do so after the fact, not knowing for sure that we have been able to take into account everything that is relevant. A combination of the relevant factors may be responsible, and so the possibilities are many. The best we can often hope for is to identify the obvious reasons for a failure and protect ourselves in the future against such a failure for those reasons.

2. **Circumstances**—In any event, whatever we may discover regarding the operator, we need to examine the situation to determine if some feature of the conditions in which the accident occurred contributed to the problem. What about the pilots for Go!Airlines? Were they trying to avoid a storm? What about an auto accident or, more dramatic,

a multiple-car accident on the Interstate? Did someone slow down suddenly to answer a cell phone? Could drivers not see because it was so foggy? We must examine all the features of the situation to try to isolate what it was about the situation that was crucial.

How different the situation is from what we designate as "normal" will play in our assessment of what needs changing. Reflectors on the side of the road or on the median strip would help when the fog is light, but perhaps make no difference when the fog is heavy. Even the most intelligent, well-trained, and highly motivated individual, working with an artifact designed to be as foolproof as possible, can have problems that cause an accident if the situation is sufficiently abnormal. Even a captain well practiced in docking a ferry can have an accident when the waves have been churned by hurricane winds. Even a driver used to ice and snow can be surprised by a patch of black ice. The conditions in which an accident occurred can be a crucial causal factor.

Assessing how much the circumstances contributed to an accident can be as difficult a matter as determining what circumstances were critical—even when no issue about the operator or the artifact arises to complicate any determination further. Hitting a patch of black ice does not always lead to an accident, for instance. Sometimes we can drive out of the skid it produces, but a failure to do that does not necessarily mean that we are somehow at fault. Some small variable in the circumstances—the patch of black ice running longer in the direction in which we are supposed to turn in such situations—can make even what seems the best response to the problem a mistake.

3. **Artifact**—Yet even with the worst of conditions, and even with what may seem operator error, we need to examine the artifact in question to determine how that may have contributed to the problem. If something about the artifact in question was a contributing factor, we should know that sooner rather than later, and waiting to determine whether the circumstances or the operator or some combination was completely at fault means delaying any fix to the artifact and risking yet another accident. So we should ask, when looking into the circumstances and any difficulties with the operator, "Was there something about the artifact that contributed to or caused the accident?" As we well know, there often is.

It is one of life's common annoyances in this technological age that the artifacts of our lives provoke, encourage, or permit errors and so create problems for us. Cell phones that appear to operate one way but operate in another and shower handles that do not turn the way they appear to. The list is long, and it takes only a query of friends and acquaintances to elicit all sorts of examples of common objects so designed as to cause accidents and provoke errors on our part. We can be in the most pristine of circumstances, highly motivated, well-trained and intelligent, and yet still have problems, provoked by the design of the artifact.

Life's accidents do not come neatly divided into those caused by the operator, those caused by circumstances, and those caused by the artifact, but we can understand how these three variables can each contribute to an accident and in some clear cases can separate out one variable as the most causally relevant for an accident.

For instance, when we hear that a small child has driven a car into some sort of obstacle, we can presume the child is highly motivated, but obviously untrained. "Operator error!" is the appropriate response. The circumstances do not matter, and the car is not to be faulted. The problem lies wholly with the operator in such a case, and clearly some operators can cause grievous harm—as with the train operator in California who sent 29 text messages while on the job, including one 22 seconds before the train he was operating ran head-on into a Union Pacific train,[16] killing 25 people. He was so distracted that he ran right through a red light and as the chair of the National Transportation Board said, he "really did not have his head in the game."[17]

Even an operator with "his head in the game," intelligent and well trained, may be unable to avoid an accident if the artifact fails. We know this because we all know that shovels break, brakes fail, electrical systems short out. The world is full of artifacts that fail, putting the best of operators, in the best of circumstances, in an accident. The artifact may have a link that finally gives out or a fault that finally shows itself when subjected to a particular stress. Things wear out, as we all know. Even something that has worked well for a very long time may suddenly fail because of age or because we put a different stress on it that is just different enough to cause it to break. We know that even the best of us, in the best of circumstances, may be unable to avoid an accident even with well-designed and properly used artifacts. The artifact can hardly be held at fault in such situations any more than we can.

But we also know that we can have faults in the original design solution or faults introduced as a design solution makes its way to realization in an artifact. A change during construction of the Citicorp Center in how the wind braces were fastened is a classic example of how a design solution failed realization. The architects had specified that the wind braces be welded, but they were bolted instead. That change made no difference to how the building would withstand winds perpendicular to the walls, the only measure required by the New York City building code. But the change put the building at risk of failure if winds of no more than 80 miles per hour hit the walls at a 45-degree angle.[18]

There are, no doubt, many mishaps that come about because of factors introduced as a design solution is being realized, and many of those may be avoided by changes in the design itself. A design solution that requires immense care in its realization is more likely to lead to a failure than one that does not.[19]

The design for the space shuttle is a case in point. When shuttle rockets blast off, they produce a twang, a vibration that moves the entire shuttle as it gathers strength to lift off. If the shuttle rocket were a single long cylinder, the twang would not harm it, but if the shuttle rocket is composed of segments, stacked on top of each other, the twang creates a separation between the top of one shuttle segment and the bottom of the next one. Hot gasses may then blow through that separation. The only way to prevent the disaster that blow-through could cause was to line each segment with O-rings resilient enough to spring back into place almost instantaneously after ignition. But the O-rings had to be put in place with incredible care since even the tiniest mistake—a hair falling upon an O-ring—could cause a catastrophic failure. It is, no doubt, true that many design solutions must incorporate features that require such incredible care in its realization in an artifact, but this solution was not the only possible solution and courted failure when others would have been less likely to fail.

So one feature of design solutions we should be wary of is the issue of complexity. The more ways in which problems can occur when going from a design solution to its realization, the more likely it is, obviously, that problems will occur. But, as indicated, I will put to one side the missteps that can arise from trying to realize a design solution so that we may concentrate on problems with the design solution itself.

An artifact may be so poorly designed as to permit or encourage errors. It may be so badly designed as to provoke errors for the most intelligent and well-trained individuals who "has his head in the game" even in the most pristine of circumstances. If the circumstances are the most favorable and the operator satisfies all the features needed for operating the artifact—intelligent, well trained, and in the zone— and does not make a mistake, then anything that goes wrong can reasonably be attributed to the artifact itself. Something about it must be the cause of what goes wrong for the best of operators, fully engaged in operating the artifact, in the best of circumstances for using it.

The harms produced by such designs can vary considerably—from minor annoyances to catastrophic disasters. It is a minor annoyance to find ourselves unable to wash our hands at a sink operated by putting our hands under the faucet. When nothing happens, we cannot tell whether we have somehow not quite got our hands in the right place, whether we are supposed to move our hands rather than merely put them in the right place, whether the mechanism is broken, or whether the sink appears to operate when our hands are placed in it but actually operates some other way, by pushing a pedal, for instance. With nothing to tell us when the mechanism is broken, we can only move to another sink and try again or leave. That sort of problem is minor, but, still, can be more than minor, and immensely annoying, when we have just changed our baby's diaper, for instance.

It is far more than an annoyance for an artifact's design to increase the likelihood of death. We shall examine several cases where a failure to think through a design solution led to catastrophic failures—from crushing a patient to death on an X-ray table to crashing an airliner and killing 159 passengers. We shall find in each case that the artifact is at fault and that the engineers who designed the artifact failed to design it properly.

We shall concentrate first on artifacts that are so badly designed that they provoke even the most intelligent and well-trained operators to make mistakes even when they are in the zone. By examining such badly flawed design solutions, we shall be able to see clearly how engineers can do harm, even without intending it. We will be able to see clearly how ethics enters into engineering practice itself. If even the best operators are led to make fatal mistakes because of an engineering design solution, the design solution is at fault.

3

The Colombia airliner crash

Newspaper accounts of an airliner crash in Colombia stated that it was the result of pilot error, but an analysis of the software involved in the automatic pilot revealed a programming mistake that would provoke an error in the most pristine conditions for even the most intelligent, well-trained, and highly motivated pilot. The example serves as an extended illustration of how a solution to a design problem can cause enormous harm.

The programming mistake was not essential for the software to perform its function, and so the engineers introduced avoidable harms. In causing unnecessary harms, they were morally at fault in choosing that design solution.

The aim is not to show that engineers should avoid such design solutions. That is obvious enough. The aim is to show that ethical considerations enter into design solutions whether, as in this worst-case scenario, great harm is produced or whether, in a best-case scenario, it is avoided.

The crash

An airliner crash in Colombia in 1996 killed "all but 4 of the 163 people on board." The plane was to land at Cali, but when the pilot turned on the autopilot by typing in "R" for the name of the navigational beacon at Cali, the plane turned slowly in an arc, heading toward Bogota, more than 90 degrees and 100 miles away from Cali. The plane crashed into the side of a mountain before the pilots were able to "figure out why the plane had turned."[20] In fact, for some time they did not even realize it was turning, it appears. They had given up control of the plane to the autopilot, it seems, and were taking care of other matters.

Nothing suggested that the weather was a factor—no wind shear, no lightning strikes, no great turbulence. So we may put to one side the circumstances as providing any significant contributing factor to the disaster. The problem is going to lie with the operator, the pilot and copilot in this case, or with something about the airplane, or some combination of those two. So what went wrong?

Each navigational beacon has a name, and the software in the autopilot uses the first letter of that name to identify the beacon. The Cali beacon is called Rozo, and so the software identifies it by the letter "R." The captain typed in "R" for the Rozo beacon. "When that letter was entered into the flight management computer, the screen responded with a list of six navigational beacons."[21] The norm is that the computer ranks beacons by distance, with the closest at the top of the list. The autopilot is programmed to fasten onto the top-ranked beacon and guide the plane into the airport without any further action by the pilot. So the pilot thought he was done when he typed in "R" for the first letter of the Cali beacon. He expected the autopilot to take over and land the plane at Cali, and that seemed to be what it was doing.

In the list of beacons the autopilot provided was Romeo, the beacon at Bogota.

The "names for the beacons at Cali and Bogota both start with R," and Romeo was one of the six closest navigational beacons. So a pilot would expect to find Romeo on the list, but would expect to find it farther down the list than Rozo. After all, the plane was about to land at Cali, and Bogota was 100 miles away. But though the norm is that beacons are listed with the nearest at the top and the farthest away at the bottom, "typing a single R takes a plane toward Bogota," the capital of Colombia.[22] That is, if a pilot types in an "R," then no matter where the plane may be trying to land, the norm no longer holds: the autopilot will list the Romeo beacon first, for Bogota, and then automatically head the plane toward Bogota to land.

The pilot expected the norm and so apparently did not notice that the autopilot's list of beacons had Romeo at the top rather than Rozo. Indeed, the pilot and copilot did not even notice that the plane was turning. They were apparently jolted alert by the air traffic controller telling them "to take a more direct approach to the Cali airport." Since the plane was on autopilot and was supposed to be going directly into the Cali airport, they were at a loss initially to figure out what was going on. The expectation that the autopilot was landing the plane at Cali apparently got in the way of their realizing that the plane was turning. It did not help that the plane was turning very slowly away from the airport, heading toward Bogota. They could not figure out what was going wrong and "spent 66 seconds trying to follow [the] air traffic controller's orders" before slamming into the side of a mountain.[23]

Operator error?

The headline in the *New York Times* story said, "Pilot's Wrong Keystroke Led To Crash, Airline Says." The airline attributed the accident, and thus the death of 159 people, to the pilot's error in locking the autopilot onto the wrong navigational beacon.

"Operator error!" is a typical corporate response to such accidents. If the operator made a mistake, it is not the fault of the company involved—except for having hired someone who could make such a mistake. So a company has an interest in blaming the operator and so trying to immunize the company from any fault and thus minimize its legal liabilities. If the operator is responsible, the artifact is not, and the company's liability for the accident is significantly lessened. We should therefore look with suspicion when an accident occurs and the relevant corporate entity says, "Operator error!" We should look with suspicion on the airline's claim in any event. In this case we have another reason for being suspicious.

The software ranks airport beacons by distance from the aircraft but shows the Romeo beacon at Bogota first when "R" is typed. Any letter except "R" will result in a ranking of beacons by their distance from the plane, the top-ranked beacon will be selected by the software, and the autopilot will guide the plane into the airport for that beacon. But typing "R" results in the Romeo beacon at Bogota being ranked at the top so that, no matter where the aircraft is, Bogota will be selected by the software, and the plane will head toward Bogota even if it is about to land at some other airport. The software creates two different sets of rankings, depending upon what letter is typed.

Saying the pilot made the "wrong keystroke," as the headline has it, hardly begins to describe what went wrong. There are only three ways in which the pilot could have made the "wrong keystroke":

First, if he had typed "T," the key to the right of "R," we would fault him for being careless. We all know how easy it is to hit the wrong key on a keyboard. We reach out our index finger to hit the "R" and hit "T" instead. Anyone who has used a computer has made this kind of mistake. It is an easy mistake to make but in the wrong situation can be deadly. But the pilot did not do that. He typed "R."

Second, we might say that the pilot made the wrong keystroke had he gotten the name of the beacon wrong, thinking it was "Tozo" instead of "Rozo" and so typing "T" rather than "R." He would then have made the wrong keystroke, but we would fault his knowledge, not his lack of care. Whereas in the first situation we would fault him for hitting the wrong key for the right name, here we would fault him for having the wrong name, even though he hit the right key for that name. But he did not have the wrong name. He typed "R," knowing, we can readily presume, that "Rozo" was the name of the beacon for Cali.

The third possibility is that he intentionally typed "R" knowing that the autopilot would pick Bogota. He could have typed "R" so that the plane would turn toward Bogota and crash. He would have made the right keystroke, given an intent to crash the plane, but the wrong keystroke had he meant to land at Cali. Yet nothing about his behavior before or after the autopilot kicked in indicates an intention that the plane fly into the mountainside. He seemed as surprised as anyone by the discovery that the plane was not heading in to land at Cali.

So saying that the pilot's wrong keystroke led to the crash is, to put it mildly, just plain false. Whether the airline knew it was false when they made that claim is another question, and we shall not pursue it. It is enough to know, for our purposes, that the airline had an interest in blaming the pilot and so trying to immunize themselves from any fault—and so minimize their legal costs should they be sued.

Predictable problems

So if it is not accurate to say that the pilot made the wrong keystroke, what are we to say? We can see that this is not a paradigmatic example where one of the three variables—the circumstances, the operator, or the artifact—is completely at fault. The circumstances we may put to one side, as indicated, but in assessing this accident, we are burdened by not being sure what the pilot knew and did not know and so cannot be sure he does not bear some responsibility.

There are only two options:

i. Either he did not know that the computer was programmed to rank Bogota first when "R" was typed, or

ii. He knew that typing "R" would rank Bogota first even though typing "R" so close to Cali would have brought Cali to the top of the ranking had the program worked the way it usually does, by ranking beacons by distance.

If (i) is true, then more training might have helped although leaving the software as it was would almost surely lead to some pilot making the mistake at some time when under stress or distracted or "not on his game" in some other way. So a reminder could have been pasted on the computer keyboard so as to be immediately visible to anyone about to key in the letter "R." Or something could be fixed to the keyboard to block the key for the letter "R," requiring that the operator do something else before "R" can be typed, thus serving as a reminder of the exception to the standard rule of ranking airports by distance. Or perhaps something could be arranged, without changing the software, that would require the operator to type some more letters to ensure that the autopilot is fastening on the correct beacon.

If (ii) is true, then he must have forgotten, or punched in "R" without thinking about it (as when we close and lock a car door by habit even as we see the keys inside), or been distracted, or whatever.

In either event, whether he knew about the oddity in the software or not, that oddity creates a problem that we can predict with a great deal of certainty would lead to something just like what happened. Instructions provide us with a rule, a procedure to be followed by so-and-so in order to achieve such-and-such a goal. The software for the computer is a set of instructions, a rule, for the computer to follow: if the operator types in "X," then such-and-such will happen. We have all been following instructions since we were little children. "Brush your teeth before you go to bed!" "Put on your clothes before you come down for breakfast!" So we know what it is like to have instructions and to follow them. We also know, however, that if the instructions are complicated by an exception, we are likely to forget. The rule that we should put our clothes on before we come down for breakfast except for the third Wednesday of the month is going to guarantee, given our nature, that we are going to forget some third Wednesday or other.

The rule we would have to teach a pilot about this autopilot software has just this form, with just that guarantee. The instructions would have to read something like this: "Type in the first letter of the beacon for the airport at which you wish to land except that, when the letter is 'R,' the autopilot will direct the plane to Bogota and if you do not wish to land at Bogota, you must type, well, something else to land at the airport you wish to land at, but we are not sure what, or land the plane on your own, without using the autopilot."

Pilots can make mistakes and are as prone to make mistakes in times of tension and stress as the rest of us. Given the necessity of instructions like that, we know that some pilot, somewhere, distracted or forgetful or whatever, will type "R" without realizing or remembering that the autopilot will direct the plane toward Bogota. We need only suppose a pilot who never flies to Bogota and has never previously flown into any airport whose beacon begins with the letter "R." A pilot could be experienced with that software, that is, and still make the mistake because there has been no occasion to remind the pilot of this oddity in the software. And a pilot not experienced in that software, but in software in similar planes, would have no reason to expect an exception to the norm. It is not likely that even the most experienced pilot would have expected the computer to choose a beacon 100 miles away when the norm is that beacons are ranked by distance, with the closest at the top of the list, and he knew that was the norm. So the software will need a way to countermand a choice—as we all know from having pushed "Send" on an email before we were truly ready to send it.

It should be noted that there is at least one other possible explanation for the accident, but one in which the onboard computer also plays a pivotal role. According to this explanation, the crew was told by the Cali controller "to report when it had passed over a radio navigation beacon called Tulua." It took the captain 90 seconds to look up the code for Tulua and "program it into the" computer, but "by the time he had done that, the plane had already crossed the beacon." Typing in the code for Tulua told the autopilot to find the beacon and pass over it, and so the plane slowly began to turn around. The captain and the first officer did not notice the turn for some time, and when they did, they turned to another computer "that directs the plane's autopilot through 'heading select' dials. They dialed in the heading they thought they were supposed to be flying," and the plane turned to the right, all the while continuing its descent. The captain took over the plane when "the ground proximity warning blared" two minutes later, but was unable to prevent the crash.[24]

Guarding against error

We know that complicated instructions with exceptions will lead us to make mistakes. We will forget the exception or forget when the exception occurs, for instance, and we can counter what we know about how we will respond to such a set of instructions in only two ways:

1. As suggested, we could create something that would warn a pilot. A sign on the autopilot might help—though it could readily be overlooked or removed by someone who does not realize its importance or fall off and not be missed. We could

put a physical block on the letter "R," as suggested—a cage
that needs to be removed before "R" can be typed. That way
the pilot has to do more than one thing in order to type "R"
and so would know that something was out of the ordinary.

Such warning devices are standard when a bad design is likely to
mislead an operator, but they leave the problem untouched. Playing
around with various devices to warn a pilot is like a physician
prescribing medication for the symptoms of a fatal disease without
bothering to find out the cause of the symptoms. The disease
continues along, untouched by the medication; just so, the bad design
continues along, untouched by the warning signs. Just as it is only a
matter of time before the disease kills the patient, so it is only a matter
of time before someone misses the warning signs and precipitates a
disastrous crash. Warning devices are not the most effective way to
handle a badly designed artifact.

2. The most effective way is to redesign the artifact so that no
warning device is needed, and that is the obvious solution
here. The software needs to be made consistent.

Either Bogota should always be ranked first whenever any letter is typed
or the beacons should always appear in the order of the distance of the
airplane from the airport. The former solution would require that a pilot
override the autopilot's choice whenever approaching any airport other
than Bogota. The latter solution would thus make more sense, requiring
nothing extra on the part of the pilot and so leaving one less opportunity
for a mistake in situations in which even the smallest of errors can cause
great harm and in which stress makes mistakes more likely.

Other solutions are possible. For instance, the beacons could be ranked in order of distance and names given to the beacons so that none start with the same letter. Or each beacon could be given a number, its coordinates, say, unique to it, and ranked in order of distance.

Other such solutions are possible because the software is an artifact, a human creation. An artifact's features, whatever they are, are those put in it—either by choice or by happenstance, and there is no necessity that any artifact be designed as it is designed. It is always possible to redesign artifacts. It is always possible to make them different from what they are, and it is always plausible to suppose that they can be made better—so that they can be manufactured more cheaply, require less energy, are less prone to failure, and, as in this case, are less likely to precipitate a catastrophic accident.

We cannot be sure we can exonerate the pilot completely in this case. Perhaps there is something he could have done had he realized what the oddity in the software could cause. If he was not aware of the oddity in the software, then perhaps he should have been, and knowing of that oddity, he should have anticipated just the sort of problem that arose. So this is not a situation where we can say with certainty that the artifact is completely at fault.

Badly designed artifact

No one can doubt that if the software had been designed so that the norm was that the closest beacon was always chosen, the crash would not have occurred—or, more carefully, would not have occurred because of the fault with the software. The feature that precipitated the crash was either chosen by the software engineer or engineers who designed it or was introduced because of some fluke in the program that they failed to catch. In either event, they are responsible—either for intentionally introducing into the software a feature that increased the likelihood of a catastrophic crash or for unintentionally designing software that permitted that feature and then not catching the oddity while testing the software (assuming that they did indeed test it).

The rush to get software out the door can often lead to a failure to test it or to test it thoroughly. Microsoft's Vista is a case in point. Quite sometime ago, long enough that I cannot now find the reference, I read a review of software for determining driving routes before GPS devices were common. The setup required that the user put in an address, but when the reviewer clicked on "Continue," he was informed that he had failed to put down the state—New York, Missouri, wherever he was. He tried to go back, but the software would not let him to do that. He tried to continue on, but the software would not let him do that. Only after crashing the program, and his computer, was he able to start the process over again. But when he went through the program again, this time paying more careful attention to its requirements, he discovered that the program did not permit an operator to put in a state of residence. Clearly the software went out the manufacturer's door without anyone there trying it out to see if it worked, or, worse, it went out even after someone tried it out and discovered it did not work.

As software gets more complex and various updates are added, the likelihood of such failures will no doubt increase. "On their first deployment to the Pacific, eight F-22 fighter jets experienced a Y2K-like total computer failure when crossing the international date line.... All onboard computer systems shut down, and the result was nearly a catastrophic loss of the aircraft. While the existence of the international date line could clearly be anticipated, the interaction of the date line with the software was not identified in testing."[25]

The crash of the Colombian airliner was catastrophic not only because it led to 159 deaths and all the consequent harm to the families and the employers of those people but also because it led to the loss of an expensive aircraft and no doubt a myriad of other harms including, presumably, an examination of the software in other aircraft, revision of the training manuals, and retraining of all other pilots—each an extensive and expensive enterprise.

Ethics enters into the heart of engineering because the autopilot software did not need to be designed the way it was and its design provoked a fatal mistake. Ethics enters via the design of the artifacts of engineering. No one is suggesting that any software engineer created the flawed software so as to mislead anyone. Moral responsibility does not depend on anyone intending to do harm. A physician who amputates the wrong limb is morally blameworthy whatever the physician's intention. The designers are morally responsible for the flaw in the software whatever their intentions. It is a moral failure not to think through how the software would create the sort of problem that led to the crash and a moral failure not to redesign the software to avoid the problem.

This accident may seem an anomaly, the oddity of the software being so unusual as to limit the lessons we can draw from it regarding engineering in general. But what drove the accident is what drives every engineering project, a set of choices about how to design a solution to a problem or set of problems. These choices are not morally neutral even when the designs chosen are themselves free of harm and innocent of any harmful effects. That just means the morally right choice was made. No engineering choice is morally neutral, that is. The solutions to design problems incorporate choices that have effects once realized in artifacts and produce more or less harm. Since we are obligated not to cause unnecessary harm, ethics is at the center of engineering, at its intellectual core, the solution to design problems.

We shall see in the next chapter how ethics enters even into the simplest of design choices—such as how to place the knobs and burners on a stove top. The aim is to illustrate with a simple example both the concept of an error-provocative design and the ways in which ethical considerations inform design solutions.

4

How artifacts can
provoke harm

It is a standard problem given to engineering students to design a stove top. There are four burners and four knobs, and their assignment is to position them all so that it is obvious which knob controls which burner. The problem is presented as an ergonomics exercise. The students are to think about how to design a stove-top layout so that a cook can use it efficiently. The first choice of which burner goes where in relation to which knob starts a chain of decisions. That first choice and each succeeding choice can cause harm, or may not. Each choice is thus an ethical choice, and the standard problem of designing a stove top thus illustrates how ethical considerations permeate design solutions, entering into what seem to be even the most mundane decisions engineers make. A mistaken configuration can lead to an error-provocative design, but we do not need that worst-case scenario to understand how ethics enters at each step in the design process.

A standard problem for engineering students is to design a stove top. The aim is to remind them that what they design will be used and that there are principles on how to design artifacts that need to be respected if what they design is to be easy to use when realized in an artifact and is not itself to cause new problems. The problem given them is to lay out the burners and knobs. The advantages of this design problem are many. The problem can be clearly stated; the solutions can be readily considered for their advantages and disadvantages; students can illustrate the problem and solutions easily by drawing stove tops; and the advantages and disadvantages—the benefits and harms—are readily perceivable. In addition, we are all relatively familiar with stove tops. We do not need any esoteric information—as with how best to place O-rings in the joints of the shuttle booster rocket segments, for instance—to make reasonable judgments about what works well and what does not, and we do not need any special equipment. A piece of paper and a pencil—and eraser—will suffice.

We shall find that the very first decision we must make in designing a stove top is ethically loaded. The decision is so obvious and straight-forward that we may not even think of it as a decision, but once we tarry and consider other choices, we can understand not only that it is a crucial decision, but that ethics enters in solving the design problem at the very beginning—not in the last chapter of the design process, but in the first.

That first decision, along with all the other decisions we must make as we work out the problem of design, is not presented as an ethical problem. It is presented as a problem in ergonomics: how are we to design stove tops so that they can be easily used? The solutions have the form of conditionals: "If I place the knobs here and the burners there, then someone using the stove top will easily see which knob controls which burner." That conditional is a factual claim, a claim about how a particular placement will affect any user, and so it may seem to have no ethical weight at all. After all, an ethical judgment is normative. It says that we ought or ought not to do something.

But when we make a judgment about how best to lay out the knobs and burners in a stove top, we are making a normative judgment. The use of "how best" brings out its normative character. We value ease of use, and so even at that simple level we are making a judgment that the design *ought* to be such-and-such if the stove top is to be easy to use.

But we value ease of use not just in and of itself, but for a further reason. We value it because if the stove top is easy to use, those using it will not easily be led into making mistakes about how to turn off a burner that has overheated a pan, for instance, where a mistake could cause a fire. So when we choose one design for the knobs and burners over another, we are thereby making a normative judgment. We are saying, "This is how we ought to design the stove top so that the cook will be able to use it easily and not cause harm." And that is to make an ethical judgment, a judgment about what we ought to do if we are not to cause harm.

So the judgments about how best to design a stove top are not just conditional claims about how someone using a stove top will be able to see which knob controls which burner. They are claims about what will minimize mistakes and so minimize harms, and in that way they are normative as well as factual claims. That is how ethics enters into design solutions at the very beginning, when we first begin to think of alternative solutions to the design problem.

The problem of designing stove tops is not presented as an ethical problem, but it is. It is a wonderful example of how ethical considerations enter into design solutions without any requirement that we think about ethics—about Kant, or Mill, or Aristotle or any of the complications of ethical theories. In solving such a design problem, we are making factual claims—"Putting the knobs this way will make it easier for a cook to see which knob controls which burner than putting them that way"—that carry an ethical punch so that the factual claim embodies an ethical judgment as well. It is not necessary, that is, that engineering students be introduced to ethical theories in order to be ethically engaged in their discipline. They are already ethically engaged.

This was the point I was making in the first chapter when I said that I was not trying to introduce ethics into engineering, but pointing out that ethical considerations are already integral to the solution of the design problems that form the intellectual heart of engineering. I am simply describing what engineers already do so as to make it clear that they are in fact making ethical decisions. As I said, that new description makes no change to what they do and does not distort it in any way, but it does bring to light a feature of what they do that has been bleached out, as I have said, by a mistaken understanding of what they do. That we can find ethical considerations entering into even the simplest of design problems, the ones first-year students get to introduce them to ergonomics, is, on my view, only to be expected.

Stove tops: How to confuse a cook

The design problem for such a simple object as a stove top is every bit as complicated as the design problem for a toothpick. The burners must be positioned in such a way that a cook can put cookware on and take cookware off each burner without a serious risk of touching a hot pan or pot on an adjacent burner. The burners must vary in size so as to fit the various sizes of preexisting cookware sufficiently well to provide enough heat to the cookware without wasting any, and the burners and knobs must be so placed that it is easy to determine which knob controls which burner.

The problem to be resolved is simple. There are usually four burners and four knobs, each controlling one burner. How can the burners and the knobs be arranged so that a cook will not be confused?

The all-too-common arrangement is this one, in Figure 4.1:

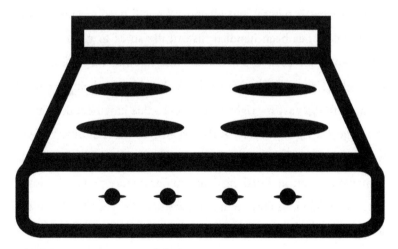

FIGURE 4.1 *Stove top with burners in a row.*

This arrangement of knobs and burners makes it anything but natural and obvious to someone using the stove which knob controls which burner.

Unfortunately, what may be natural and obvious to some regarding any particular design may not be natural or obvious to some others. What is natural for a right-hander need not be natural at all for a left-hander. Just think of how door knobs turn. A clockwise motion is natural for a right-hander, but requires an effort for a left-hander. Another variable that can matter to what seems natural are the habits we have come to have because of some previous solution to a particular problem. The keyboard arrangement on our computer seems natural—as anyone who has tried to use a different keyboard configuration can attest—but that is only because we are all used to the now-standard keyboard arrangement, the QWERTY. Arranging knobs on a stove top so that which knob controls which burner is natural and obvious requires more than simply following our own sense of what is natural and obvious. We will need to look at how the problem has been handled historically so as to get a sense of what people are used to and so now find natural. We shall need to determine if the past solution has been biased by, say, being arranged to make things easy for right-handers—a value judgment that favors the majority, right-handers, at the expense of the minority. We will need to look at what others find natural and obvious, comparing their responses to ours and making sure that our own responses, however natural and obvious they seem to us, are not idiosyncratic. The list is long, and so the question is not as simple as it looks.

It is also of no small importance. Suppose we arrange the knobs randomly. There are twenty-four possible combinations. The first knob on the left could control any of the four burners, the second any of the three, the third either of the two remaining, and the last knob the last remaining burner. We can imagine how confusing such random combinations would be by supposing ourselves standing in front of the stove, trying to figure out which knob turns on the burner that has the pan with our eggs to scramble. First one on the left? Maybe. Maybe not. Second one on the left? If the distribution is random, the only way to answer the question of which knob controls which burner is to experiment and see what happens.

Because a random distribution is going to be anything but natural and obvious to anyone, we are surely less likely to remember the next time we use the stove which knob controls that burner, and if the knobs and burners are randomly distributed on every stove top, we cannot carry any useful information from our experiences with one stove top to any other.

Distributing the knobs randomly would certainly confuse a cook. We could also easily confuse a cook by putting the knobs for the right-hand burners on the left-hand side and the knobs for the left-hand burners on the right-hand side. Someone trying to turn on the front right-hand burner would reach—naturally—for one of the two knobs on the right-hand side, not knowing which of those two controlled the right-hand front burner.

But whichever was chosen, the burner would not go on. Instead, a burner on the left-hand side would go on. If the stove was gas, the flame would make it clear that a right-hand knob controlled a left-hand burner. If the stove was electric, it might take a bit of time before the burner heated up sufficiently to allow the person to see or feel the problem. So now what should the person do? Try the other knob on the right-hand side? Move to the knobs on the left-hand side and see if either of those worked to turn on the right-hand front burner? Nothing about the information so far gained gives the person a clue about what next to try. Experience does not give us any hint.

The only way to proceed, now that the normal expectation of the right-hand knobs controlling the right-hand burners has been stymied, is, again, to experiment by trying each knob in turn and seeing what knob controls what burner. Once the pattern is uncovered, unfortunately, nothing about it will ensure that the person will remember it the next time regarding this stove top, and so a chart will need to be made and taped to the stove top or nearby to remind the person of the arrangement of knobs to burners. And if we do not tape instructions where the next person using the stove can see them, we will leave every other operator in just the fix we found ourselves in—without a clue, judging from the relative positions of the knobs and burners, which knob controls which burner.

Again, we cannot necessarily carry the information we have gained from our experiment with this stove top to any other stove top. The oddity of the arrangement ought to make us leery of the next stove top we encounter perhaps working in the same way. So one result of having the left-hand knobs control the right-hand burners, and vice versa, is that we will not be able to learn from experience. Everything old will be new again.

All we need to do is to imagine many artifacts in our lives with randomized controls. The lights and switches in a house could be so arranged that the switches do not correlate with the lights nearby but with some distant light in some distant room. Sometimes when we turn on the light switch in our car the lights go on and sometimes the windshield wipers, and sometimes the windshield wiper control opens a window and sometimes the radio. That we have patterns of correlated controls is not a trivial matter, in other words, but makes it possible for us to have habits of behavior and so live a life without having constantly to learn anew everything we normally do.

We could complicate matters even more by adding some electronics to the stove top and programming the relations between the knobs and burners so that which knob controls which burner changes randomly. Then no chart would help, and we would approach any stove top with trepidation rather than any expectation about which knob controls which burner. Just imagine a computer or house lights organized or, rather, disorganized in such a manner!

Burners and knobs: The first design choice

The standard way of configuring burners divides them into left-hand and right-hand burners, but there being left-hand and right-hand burners requires a prior decision. Indeed, a classic study by Chapanis and Lindenbaum of "preferred locations of controls for burners on stove tops" was designed with that decision already made. The options presented to the subjects of their study were already constrained by a prior decision that the left-hand knobs control the left-hand burners and the right-hand knobs control the right-hand burners. Having made that decision, the two thus examined not twenty-four, but "four alternative layouts of the burners."[26]

That decision may have seemed so obvious that it may not have occurred to them to study what would happen were all twenty-four possible combinations considered. But however obvious the solution may be, it still requires that a decision be made, and we can now see its importance. Failing to make that decision will stymie anyone trying to use the stove top. The far-right knob might control any of the four burners, and the only way to determine which knob is to turn it on and see what happens. Ditto for the far-left knob. Nothing about it or its location gives a clue to anyone trying to use the stove top what it turns on. Unless someone draws up a chart for themselves and others or has an incredibly good memory, each approach to the stove top is an approach to foreign territory where one has to make out the meaning of the road signs. Without some organizing principle such as the right knobs controlling the right burners and the left ones controlling the left burners, any arrangement would seem arbitrary.

Some harms coming from any such arrangement would presumably not be too serious. Those using the stove top would be confused and become irritated, and at some point someone will likely leave a burner on, being called away by a phone call, perhaps, before realizing that though a right-hand burner did not go on, a left-hand one did. That harm would be serious because it could be dangerous. We need only imagine something overflowing from a cooking pan and the cook being unable, under the stress, to find the right knob to turn the burner off and so starting a fire—just like the person who "irreparably damaged" the "internal mechanism" of that toaster by pushing the lever when the bread started to burn.

No doubt the company manufacturing such stove tops would eventually face a lawsuit because of some serious accident brought about by the irritating arrangement of knobs. Just as bad for the company would be that once it got a reputation for such a wildly unnatural arrangement of its control knobs, it would lose a great deal of business—and presumably fire the engineer who thought such an arrangement acceptable. Who wants to hassle with such a problem when you do not have to?

The obvious solution, then, is to let the two knobs on the left side control the two burners on the left side and the two knobs on the right side control the two burners on that side. Left knobs control left burners, and right knobs control right burners. So far so good, then, but how should the knobs within each set of two be arranged? That raises a second set of ethical decisions.

Burners and knobs: The second design choice

Let us call the burners, from left to right, and front to back, left front (LF), left back (LB), right front (RF), and right back (RB). We could have the left knobs of each pair control the front burners and the right knobs control the back—in the order LF, LB, RF, and RB. Unfortunately, that is only one of four different possible arrangements if the left-hand knobs control the left-hand burners and the right-hand knobs control the right-hand burners. The knobs could be arranged so that the middle two knobs control the front burners and the outer two control the back burners—LB, LF, RF, RB. Or we could have the outer knobs control the front burners and the inner two knobs control the back burners—LF, LB, RB, RF. Or we could have the left knobs of each pair control the back burners and the right knobs control the front burners—LB, LF, RB, RF.

None of these arrangements is any more natural than any other. As all of us who have used stove tops know, we cannot tell just by looking at the knobs and the burners how the knobs and burners relate to each other. We all know this from our experiences in having messed up using stoves and from manufacturers putting directions next to each knob indicating which burner it controls. The directions on my current stove top spell out the words in capital letters: LEFT FRONT, and they are placed next to a set of four small circles, to represent the burners, with the circle for the relevant burner filled in. The stove top thus has written directions, with a pictorial representation—two different ways to instruct users. The directions are immediately in front of the knobs, which are set at a slant to the stove top, and so whenever I reach to turn on a knob, I can see the directions—although they are too small to read except by getting closer. That is a somewhat effective solution to the problem, but it is a solution necessitated by an arrangement of burners and knobs that calls for directions because nothing about the arrangement itself gives someone using the stove top any directions. Adding directions about which knob controls which burner is like adding warning devices for the odd autopilot software. No directions would be needed if we could arrange knobs and burners so that the cook could tell just by looking at the arrangement which knob controlled which burner.

Could the burners and knobs on a stove top be arranged to be user-friendly? The answer is not to add instructions to a design, but to figure out how to rearrange the burners and knobs so that the proper way of using them leaps out at a user, making it all but impossible to make a mistake.

What leaps out at us may not leap out for everyone. So we would need to test any potential solution to determine whether it is idiosyncratic and whether it accords, or not, with the embedded habits of those who use the artifact. There is little sense in designing an artifact, no matter how elegant the design solution, that can only be operated by doing what runs against the grain of what habits have led us to expect. That does not mean being tied down to an old way of doing things. It does mean taking those old ways into account in the transition to something new.

The transition from door latches to door knobs in the 1820s and 1830s was proceeded in some cases by an intermediate stage where the door still latched, and the latch was still operated by pulling up on the lever that came through the door, but the latch was underneath a knob. Instead of grabbing hold of a handle and pushing down on a lever, we reach for a knob and pull up on the lever. The knob does not turn on this transitional door handle, but is screwed onto the door and is used only as a handle to pull open the door. I do not provide this as a good example of taking into consideration people's habits about unlatching a door while introducing a new mode of entry, door knobs, but this combination of latch and knob illustrates an attempt to accommodate past practice.

An engineer thus needs to consider more than simply numbers— having four knobs for four burners, say, rather than three or five. We have assumed four knobs for the four burners, but it takes a decision to reach that point. It may seem odd to suggest that a knob might control two different burners, but having a single knob or switch do multiple tasks is not all that unusual. The light switch on some Dodge Caravans, to mention just one example, turns on the lights by being turned to the right, clockwise, and turns on the running lights by being pulled out. Nothing about the knob tells a driver that it serves the second function as well as the first. A driver needs to read the instruction manual to discover that the single switch serves a double function. There could also be two controls for a single burner, a switch to turn it on, say, and a knob to regulate how hot it gets. That might be inconvenient, but there might be reasons for such a design solution.

In any event, no matter what design choices seem best, an engineer needs to consider features of human psychology. What do human beings, as human beings, find natural in arranging things—if anything? An engineer needs to consider the history of technology.

What habits do humans have as a result of previous solutions to this problem and, just as important, as a result of solutions to similar problems? History? Psychology? Engineers need more than mathematics and physics in order to succeed at the core intellectual endeavor of engineering, the solution to design problems.

So how can we place the burners and knobs to preclude mistakes? The usual solution is to shift the back burners in one direction, shift the front burners in the other, and line up the knobs with the burners so that, as in Figure 4.2, anyone can tell which knob turns on which burner by just looking at the knobs:

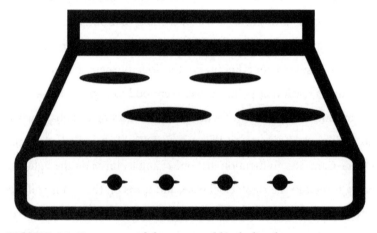

FIGURE 4.2 *Stove top with burners and knobs lined up.*

Here the order of the knobs corresponds to the order of the burners. The knob on the far left appears almost directly in front of the left back burner, the second knob from the left appears directly in front of the left front burner, and so on down the line.

My current stove top has an arrangement like that in Figure 4.1, and, as I said, it compensates for the confusion the arrangement causes by having little symbols on the knobs that tell you which burner the knob controls. What is wrong with the arrangement of burners and knobs on my current stove top is that nothing about the arrangement gives me a clue about which knob controls which burner. The advantage of the arrangement in Figure 4.2 is that you can see which knob controls which burner. You have a visual clue on which to proceed, and, as shown in Figure 4.3, we know—without really thinking about the issue at all—which knob controls which burner:

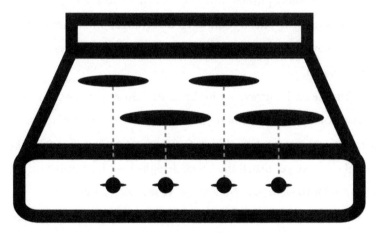

FIGURE 4.3 *How the stove top in Figure 4.2 appears to work.*

The visual clues we get from the arrangement match the actual arrangement of knobs and burners, and we do not need any additional information to get things right. So there is no need for warning devices of the sort I find on my stove top. No symbols on the knobs are necessary. Anyone new to the stove can see how it operates, and anyone who has used the stove will know instantly how to use it again and any similar stove. This is an elegantly simple solution to the problem with which we began.

Burners and knobs: Variant design choices

It is easy to imagine variants of this sort of solution. The front burners could be shifted to the left and the back burners to the right, or, less obviously, we could arrange the burners in a circle with both back burners pushed toward each other and the front burners pushed away. The knobs could then be arranged in a similar circle so that they line up with the burners. Figure 4.4 illustrates such an arrangement.

With the back burners set in the center, closer together, the front burners off to either side in the front, and the knobs set in the same arrangement, we have a solution that appears foolproof. Again, the visual clues we get match the actual arrangement for turning on the burners. Or so we would think when we saw such a stove.

I have a friend whose very expensive stove top is arranged in just this way. I was watching him cook dinner one evening. He had lobster and angel-hair pasta in separate large pots on the back burners. I was

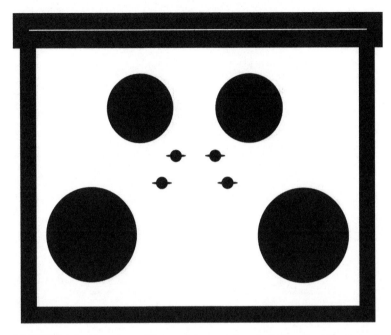

FIGURE 4.4 *Stove top with rear burners close together, front burners to the right and the left, and knobs on the top of the stove.*

sitting on a stool, drinking wine, listening to kids in the background. The stove top is gas, and I could see the flames from where I sat. I thus saw that when the timer rang for the angel-hair pasta—90 seconds to be just right—my friend turned off the burner for the lobster, leaving the pasta boiling away. I pointed it out. He said, "I'm so stupid! I do this all the time!" He turned the burner for the pasta off and then turned the burner for the lobster back on. I went over to look, curious about what could have gone wrong.

It did not look as though anyone could make a mistake here of the sort my friend made. How could he have turned off the wrong burner when it is so obvious which is which? Figure 4.5 shows how the burners and knobs appeared to be related:

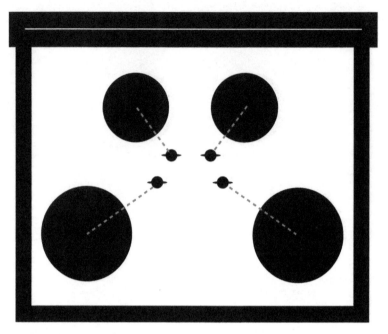

FIGURE 4.5 *How the stove top in Figure 4.4 appears to work.*

The knobs mirror the arrangement of burners, and the information we get from looking at the two arrangements of knobs and burners seems to make it obvious which knob controls which burner.

If the pasta was on the right-back burner, it looks as though all my friend had to do was to turn the right-back knob. To turn off the burner with the lobster pot by mistake, my friend, being right-handed, would have had to reach across the right-back knob and turn the left-back burner knob. That would be possible, presumably—anything is—but he complained about doing it "all the time," and making that mistake all the time does not seem likely for someone like my friend who prides himself on his cooking and usually concentrates on what he is doing.

When I asked him to show me what he had done, he pointed to the knob on the right—the one that appears to operate the right-hand back burner, the one with the pasta pot on it. "I turned that knob," he said. Now I really was puzzled. Why did turning that knob not turn off the right-back burner? What was going on here?

After some investigation, my friend and I discovered that whoever had designed, or assembled, the stove top had reversed the back two burner knobs, as in Figure 4.6. The stove top worked like this:

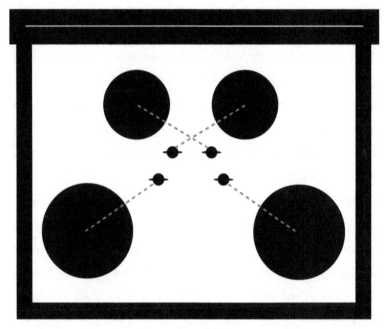

FIGURE 4.6 *How the stove top in Figure 4.4 really works.*

No wonder my friend made these mistakes all the time! He was getting a clear signal from the position of the knobs about what he ought to be doing, and he did it. But he was getting the wrong information. He had originally turned off what *appeared* to be the knob for the burner for the pasta and left on what *appeared* to be the knob for the lobster, but since the knobs were reversed, he left on the burner for the pasta and turned off the burner for the lobster.

His initial reaction was that he was at fault. Operator error! We often see this response in regard to major accidents. A major meltdown averted at a nuclear plant! The cause of the problem? Operator error. A train wreck in New Jersey. The cause of the problem? The engineer failed to slow down when the signal told him he should. Operator error. An airplane accident in Colombia? Operator error. When Ford vehicles with automatic transmissions "popped out of park and into reverse," Ford's reaction, and the reaction of the National Highway Traffic Safety Administration (NHTSA), was to blame the driver for not putting the "car fully in park." Toyota and NHTSA made the same move of blaming the operator with their recent responses to problems Toyota drivers have with accelerators revving up. The problem, they claimed, was that drivers were pushing the floor mats up under the accelerator pedal.

The executive director of the Center for Auto Safety, Clarence Ditlow, said, "Both Toyota's and Ford's reaction is to blame the issue on driver error. In the '80s, they said the driver didn't put the car fully in park—they left it in neutral or what have you. In Toyota's case, it's the floor mat's fault. The manufacturers want to avoid a costly engineering recall. For Toyota, any recall that goes beyond the floor mat will be very expensive." Toyota's problems with accelerators revving up date back at least to 2003, and NHTSA's response to complaints about sudden acceleration was that the problem was "pedal misapplication,…blaming the drivers for hitting the gas instead of the brakes."[27] "Operator error!"[28]

My friend's initial reaction, however, was not just that he had made a mistake but that the whole thing was his fault. As he said, "I'm so stupid!" So one ironic consequence of the error-provocative design of his stove top was that he blamed himself for making such a "stupid" mistake. Instead of wondering what it was about the stove top that made him repeat the mistake, he simply threw up his hands and confessed to his stupidity. The irony, of course, is that he blamed himself for what the stove top's layout made it all too obvious that he ought to do.

The fault in my friend's kitchen was not with my friend, that is, but with the design of the stove top. Perhaps my friend should have learned by this time, overcoming each time he used the stove the information he received visually from its layout and remembering that the knobs do not work the way they appear to work. But however one parcels out responsibility here, the stove top ought to get a large share. It looks to be designed to avoid the very problem its arrangement of burners and knobs create! No wonder he kept forgetting.

My parents' stove top

My friend's problem with his stove top hit a responsive chord with me. Every time I went to visit my parents, I made a mistake with their stove. Their stove top was laid out so that the burners in the back were shifted to the left, with the burners in the front shifted to the right and the knobs lined up precisely with the burners as we saw in Figure 4.2—the usual solution to the stove-top problem. As Figure 4.3 illustrated, it ought to be obvious how to use such a stove top.

My parents had no coffeemaker, and so I would have to heat water for my coffee on the stove. I always used to make the mistake of putting water for coffee on the back left burner and then turning on the wrong burner. If I was lucky, I would walk back into the kitchen after showering and find the left-front burner blazing away while the pot for the coffee sat, still cold, in the back. If I was unlucky, my mother would have walked in before me. The water would now be heating up because she had turned off the burner in front and turned on the burner in the back, but my mother would give me that look only mothers can give: how did you ever survive this long?

What was the problem? Why could I not remember? Why did I keep making the same mistake? I finally figured it out. Figure 4.7 shows how the knobs turn on the burners:

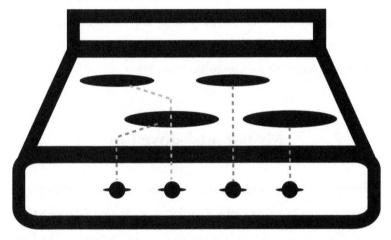

FIGURE 4.7 *My parents' stove top—how it really worked.*

I got mixed signals from the stove's design and from the placement of the knobs for the right-hand burners. The offset pattern of the burners suggests a corresponding order to the control buttons—LB, LF, RB, RF. But the left-hand knob turned on the front burner, not the back. It did exactly the opposite of what the offset pattern of burners led me to expect and exactly the opposite of what using the right-hand knobs would lead me to expect. It is not that the design is neutral between competing interpretations and that I made the wrong interpretation. The design provoked a mistake. It provoked me to do exactly the opposite of what I ought to do to heat up the coffee water just as the design of my friend's stove top provoked him to do the opposite of what he should have been doing.

My mother was so used to using the stove top that when I remarked on how the control knobs were in the wrong place, she failed to see a problem. There were no instructions on her stove top, as on my current one, but she had so trained herself that using the correct set of knobs had apparently become second nature to her, a habit so deeply embedded that she did not even realize it was a habit and that others, unused to the stove top, might have a problem. She thought the problem was with me, not with the stove top—"Operator error!" The most likely explanation for the problem with her stove top was that the wires connecting the burners to the knobs were switched when installed. The problem with my friend's stove top is harder to explain. The usual method of control for the gas to a burner is via the knob. A small gas pipe comes in at the knob, and turning the knob opens that pipe and directs the gas down a metal tube to the particular burner controlled by that knob. The lengths of the metal tubes coming from the knobs to the burners will vary in length depending on the distances from the knobs to the burners. To reverse the knob controls for a gas stove top thus requires switching the metal tubes.

We can understand how someone might mix up the wires on my mother's stove top. It requires less work to make the wires all the same length and tuck any extra wire into a corner somewhere, and with all the wires the same length while working under the top without seeing how the knobs and burners are supposed to be arranged, a worker might well reverse the connections. Her stove-top arrangement may have been a one-off wiring mistake. But the tubes cannot be switched in that way. Had the tubes for the back burners been made just long enough to reach the knobs the design indicates they were meant to reach, they could not be switched accidentally or intentionally. They would be just the length necessary to go from the center of each back burner to its corresponding knob and so too short to be connected to the wrong knobs.

What could have happened? One unlikely possibility is that the stove top was designed in just that way. But even a perverse sense of humor hardly suffices to explain a design choice that would lead to such confusion and potentially cause great harm. Another unlikely possibility is that the engineer made an error in putting the design solution to paper, somehow switching the relationship between the two back knobs and the burners in laying out in detail how the burners and tubes and knobs were to be arranged.

Yet it is more likely that some misstep occurred between the execution of the details of the design and its final realization in a stove top. A worker might have run out of shorter tubes, and, as luck would have it, the tubes for the front burners were just the right length to fit the back burners if the knobs were switched. The tubes would have to crisscross each other, and so it ought to be fairly obvious to any inspector, or any worker, that the wrong length tubes had been used. Presumably seeing the tubes crisscross would signal that things were not as they should be. Another possibility is that the manufacturer judged it less expensive to make all the tubes the same length, and so ruin the elegant design solution, than to make tubes of different lengths to save the design.

In any event, whatever the cause of the switch, something trumped the value of insuring that whoever uses the stove is not misled into making a mistake and thus causing harm. As we know, some resulting harms may not be great. Those who use the stove top will either habituate themselves to turning the correct knobs or, like my friend, continually berate themselves for being "stupid." But the design clearly provokes confusion, irritation, and errors. The errors in question that I have mentioned—overcooked pasta and my own embarrassment in front of my mother—hardly rise to the level of ethical concern, but it is not hard to imagine someone accidentally causing a fire by turning on the wrong burner without realizing it or allowing a fire to spread when a pot flares up by not being able to find the proper knob to turn off the burner quickly enough.

So designing a stove top that provokes errors on the part of those who use it is wrong. A bad design can cause great harm, and, after all, even one person's annoyance may harm someone else. That is why doors that are difficult to open even for those with sufficient strength are changed so that those without sufficient strength—some of the elderly, say—can open them. And as we know from the Colombia airline disaster, design errors can be fatal.

Ethics and design

One advantage of working through in such detail the design problem for an easy-to-use stove top is that the misaligned knobs and burners jump out from the visual images. It is difficult not to feel sympathy for my friend with his expensive but flawed stove top.

When you look at the arrangement of knobs and burners, it is difficult to imagine how anything could be made more clear. We see the arrangement and, without any more thought, reach for what for all the world looks to be the correct knob.

The point of the arrangement is to bypass any need for thought about which knob controls which burner. A cook can concentrate on cooking, not on how to turn the burners on and off. My friend was "told" by the arrangement which knob controls which burner and, having done what he was told, without thinking about it, discovered that he turned off the wrong burner. He was caught up short, as we all would be, and wondered what had gone wrong. "How could I have misread what the arrangement of the stove top told me? Stupid me!" But as the visual images make clear, the stove top is at fault.

That stove top would provoke errors in anyone who used it, no matter how intelligent, highly trained, and "in the game" the person is. A cook trained to the eccentricities of this stove top might, like my mother, become so habituated to its odd arrangement of knobs and burners as not to make the mistakes my friend had committed, and yet might still find themselves sometimes wondering if they had made a mistake, the power of what they would see in looking at the stove top putting a question to what they have become habituated into doing. They might always have to overcome what the arrangement says about which knob controls which burner. Such can be the power of that arrangement on our actions. It is sad, and ironic, that the visual arrangement of the knobs and burners on my friend's stove top appears to have been designed to avoid just the sort of error the actual arrangement caused.

So one thing we have gained from working through the stove-top problem is a paradigmatic example of an error-provocative design. Keeping that example in mind will help as we come to examine many an example of engineers having produced such designs.

This is not to suggest that the problems with my mother's stove top and my friend's are the fault of any engineer choosing a bad design. As suggested, the actual problems created by the particular arrangements no doubt came about because of bad wiring and bad judgment on the part of someone other than the engineers. These examples are meant to illustrate the concept of an error-provocative design—just how a design can provoke errors on the part of even the best of operators—as well as suggest that we might be better served by having the designing engineers follow through on how their design solution is realized in an artifact.

In any event, once the concept of an error-provocative design is clear, with such easily remembered examples, we can proceed to consider other examples where it is not so obvious that anyone was involved other than the engineer solving a design problem. Indeed, that the problems with the stove tops my friend and mother had could have been introduced through design choices by engineers is all we need to make the point that ethical considerations are integral to the intellectual core of engineering.

But our examination of the stove-top problem has also told us a great deal about solving design problems:

a. **Design problems underdetermine solutions.** As the examples of the toothpick and the stove-top design problems both show, design problems leave plenty of conceptual space for alternative solutions. There are many ways to design toothpicks, that is, just as there are many ways to design stove tops. An engineering design problem is not like a math problem in which the premises determine the conclusion. It is not wholly quantitative, that is, although it is constrained by quantitative considerations. If a design problem determined its solution the way 2 + 2 determines its conclusion, there would be but one solution. That the constraints articulated in the design problem do not determine one particular conclusion means there is conceptual space for the imagination and creativity of engineers between an engineering problem and its solution. That conceptual space allows us to credit engineers with more than the skill to calculate. In the best of cases, with a brilliantly elegant solution to a design problem, they deserve praise for a creative imagination, a difficult talent to realize and perhaps even more so for someone trained into thinking quantitatively.

b. **A design problem involves a complex of decisions in which any one decision can both constrain and open up new possibilities.** Once we choose to put burners in sets of two to the right and left, we constrain ourselves to only two different ways of arranging the knobs on either side—unless, that is, we really want to produce an error-provocative design. Similarly, if we choose a design solution for a toothpick that emphasizes breaking it after its use so that it cannot be used again, something like the Japanese solution seems mandated. We have not emphasized the cascading effect of decisions, but it is worth noting if only because we can readily find ourselves caught in a cascade, trying to figure a way out of the problem we have now come to have, without realizing that a prior decision that could be revisited created the cascade that is now causing us problems. Decisions have their effects, and one effect is that we end up thinking in a certain way because a decision closes off some possible solutions so we no longer consider them and opens up new possibilities and problems that also constrain our thought. We need to remember, in working through a design problem, the decisions we make as we go along so that we can revisit them and reconsider should we run into unforeseen problems that we cannot readily solve. We may have led ourselves down a garden path, without realizing it, and so become lost without realizing we need to retrace our steps.

c. **The choices we make in picking design problems reflect values.** How design problems are chosen, how they are ranked as worth solving, and even how we go about trying to figure out how to solve them—all these are value-laden enterprises, as value-laden as the ways in which a design problem is conceived and stated. In concentrating upon design solutions, we are leaving to one side concerns about design problems themselves, but they provide a rich source of issues for anyone concerned to see how ethical considerations permeate the core of engineering.

d. **The choices we make in solving a design problem reflect values.** They can reflect the worst of values, we know, because we know engineers can solve a design problem in a way that provokes errors. Such a solution would involve a particular configuration of values, ones that we should eschew because of the harms the design choice is likely to produce once it is realized in an artifact. An engineer could avoid such a solution and that unfortunate configuration of values by choosing some other solution and thus some other configuration. It is a mistake, that is, to think that the intellectual core of engineering is wholly consumed by quantitative considerations. It would be far more accurate to say that the intellectual core of engineering reflects value considerations. Whatever design an engineer chooses, it will reflect a set of values and be ranked the best solution to further that set, the worst, or somewhere in between— brilliant, mediocre, acceptable, or some other less than sterling choice. An engineer is effectively choosing a set of values in choosing one design solution over another. A recent solution to the problem of opening cans has produced a can opener that leaves no sharp edges, cutting through the can below the edge at the top and bending back the edges on the side and the top as it cuts through. Safety is a value, and emphasizing it in regard to this artifact has fundamentally altered the way a can opener works. The design solution that led to this can opener reflects a value choice and, in this case, a good one that trumped whatever other considerations might have mitigated against it.

e. **Some of these values are ethical values.** As we saw, ethical considerations enter engineering by the very act of engineers choosing a design solution, and they enter as well when engineers take the idea they have for solving a design problem and execute it. Putting a design solution to paper or computer in the clearest possible way is a separate task from solving the design problem and one equally prone to errors. It is an engineer's responsibility to make absolutely clear what needs to be done to whoever is to take the design solution and produce an artifact. Engineers need not be driven by ethical considerations in making design choices or in executing them for those choices to embody ethical values. They need not make explicit, even to themselves, what ethical values they are achieving through their particular design choice or even whether they are achieving any ethical values at all. However they arrive at a choice, that is, whether they choose an error-provocative solution or one that solves the original design problem elegantly, without unnecessary harms, that choice reflects ethical values and will come to embody ethical values in at least two different ways, reflecting, as we shall see, three different kinds of reasons for how ethical considerations enter into the core of engineering:

i.   Although the design choice itself reflects ethical values, as we have argued, the values can perhaps best be seen once the design is realized in an artifact: the artifact itself will embody those values. Its creation will cause more or less harm, depending, for instance, on how it is manufactured and how the resources necessary for its creation are produced. We have not raised this issue in any of the examples we have examined so far, but the point should be obvious. Choosing a design that uses fewer resources or fewer harmful resources than some other design is not an ethically neutral decision. Choosing a design that can be realized in an artifact with less expenditure of energy than another design is not ethically neutral. Choosing a design that allows for remanufacture and/or recycling is not ethically neutral. Perhaps the manufacturer of my friend's stove top decided that it was more cost-efficient to make a single-length tube. That was not an ethically neutral decision. The presumption should be that all decisions about a design carry implications that are ethically loaded.

ii. Once a design solution has been chosen, an engineer is to make use of the rules of skill essential to the profession to execute the solution in such a way as to be clear—as foolproof as possible—so that the solution can find its way to become an artifact. Engineers have a moral obligation, at a minimum, to use the rules of skill of the profession competently, and a failure to use them properly is a moral failure.

iii. Once the artifact is introduced into the world, it will have causal effects. These effects are likely to be a mix of good, bad, and indifferent, and if the total set of effects is more harmful than it need be, it would have been morally preferable to have chosen a different design solution. We need only think of such examples as the ignition switch on some GM cars. Their introduction into the world had effects. Some were minor—such as removing everything from your key chain except the car key to keep the weight

low and minimize the risk of its being knocked and so
dislodged by a knee, for instance, while driving. And some
effects were major—such as the loss of those lives that
need not have been lost had the proper part been used.[29]
Introducing any artifact into the world has effects, that is,
and an error-provocative design guarantees that some of
the effects will not be what an operator wants—whether it
causes great harm or not. So choosing an error-provocative
design is, in that sense, the worst moral solution to a
design problem. It can obviously cause great harm—loss of
life, for instance.

f.   **What an engineer chooses as a design solution will reflect
    on the engineer's character and moral values.** Though we
    have not raised this issue in the discussion so far, we need
    only go through the possibilities to see how a choice reflects
    back on the engineer. Assuming that we are free to choose as
    our design solution whatever among the range of possibilities
    we wish, we would have a difficult time explaining why,
    among all the possible design solutions, we chose the worst.
    What would we possibly say? "Why should I care?" Or,
    "What's it to you? I'm evil." And if we did not choose the best
    design solution, we would have to say something that makes
    our attitude and level of competence clear: "I'm satisfied
    with being mediocre." We would, that is, have a difficult time
    explaining why we chose a design solution that, once realized
    in an artifact, caused harm when it did not have to, polluted
    when it did not have to, frustrated users when it did not have

to, and so on. So we can look at a design solution and reverse engineer, as it were, the decisions that led to it, and if a pattern exists, the relevant features of the character that produced it.

In examining design solutions to the stove-top problem we have uncovered not only that engineers ought to avoid error-provocative designs, but also that they ought to strive for the best design solution. We will generally put that moral imperative to one side, it being enough to show how ethics enters engineering that engineers ought to avoid causing unnecessary harm. But, as we have said, engineers are no different than anyone else trying to solve a design problem. A poet makes design choices in crafting a poem, and the aim is the same: solve the design problem in the best way possible. Just as engineers ought to avoid such mistakes as producing an error-provocative design, a poet ought not choose a word that resonates or is ambiguous in a way that undercuts rather than furthers the point of the poem. Just as poets ought not to settle for mediocre choices, but strive for just the right word or phrase or rhythm, so engineers ought not to settle for mediocre design solutions, but strive for the best.

In the case of the stove top, ethics enters as soon as an engineer considers whether the right-hand knobs should control the right-hand burners and the left-hand knobs should control the left-hand burners. Indeed, ethics enters even if an engineer does not consider the issue, but simply assumes an answer. Ethics enters, that is, whether the arrangement of knobs to burners is the result of intentional action or not. It is not the intention that causes harm, but the arrangement, and it is not the engineer's intention that matters, but the failure to try to minimize the harms that will come from using an error-provocative design. We need to lay out that argument regarding the irrelevance of intentions in much more detail, however, if engineers are to be convinced that in choosing a particular design solution to an engineering problem, they are engaged in ethical, or unethical, behavior whether they realize it or not.

5

Moral responsibility: How ethics is integral to engineering

We learn early on that "I didn't mean to..." can relieve us of responsibility, and so we may come to think that we can only be held morally responsible if we intend to cause harm. But that is not true.

Engineers take on special moral relations when they hold an additional position—wearing another hat as an employee, contractor, or manager, say. We would obviously hold software engineers morally responsible for intentionally creating a flaw in software, but we also hold them morally responsible for creating flawed software even without any intention to cause harm. They have failed to fulfill an obligation to their employer.

The failure to do what they were hired to do or what their employer had assigned them to do is an external moral failure because it is a failure they have as contractors or employees who happen to be engineers. It is not internal, not one they have just because they are engineers.

But engineers can also be faulted for failing to execute the rules of skill they must learn to become engineers. We can readily imagine an evil genius of an engineer who intentionally causes harm by failing to follow the rules of the profession, producing design solutions for artifacts that will inevitably fail. But engineers do not need to be evil to be faulted morally for failing to execute the rules of skill they ought to have learned in becoming engineers. They can fail to be competent practitioners, and we will then hold them morally at fault even without any intent on their part to cause harm. These sorts of moral failures are internal to engineering since they relate to what engineers must do as engineers.

One issue we have left hanging is the role of intention in being held morally responsible. Consider again that airliner crash in Colombia and the software problem that contributed to it. No one has suggested that the software engineers purposefully introduced the flaw into the software, and someone may ask, "How can anyone hold them morally responsible when they did not intend to cause any harm?" If I accidentally bump into someone and say, sincerely, "Oh, sorry, I tripped," that generally gets me off the moral hook. I could be held morally responsible if I were unusually careless when I walked, horsing around and not paying any attention to those around me. Any person I then bumped could well chastise me for being so careless. But with no intent on my part to bump into the person, I am generally not morally at fault, and the person would be morally wrong to hold me responsible for the bump.

We learn about the relevance of intention very early in our lives. Listen to two children having a spat, or to a parent addressing a child after some misadventure, and you are bound to hear, "I didn't mean to …", followed by whatever it was that the child did for which he or she is being admonished. The child is denying any intention to cause harm, the lesson having been learned at a very young age that intention is what seems to make the difference between being held responsible and being let off.

We carry this lesson through our lives even to the point of thinking that intentions are all that matter, that acting "in good faith," without intending to cause harm, is enough to get one off the moral hook.[30] This view so permeates our lives that, I suspect, we do not even think about using it in situations in which acting "in good faith," even without an intent to cause harm, would hardly suffice to justify the behavior in question. For example, when it investigated the CIA agents who broke into the computers of the Committee investigating whether the CIA was guilty of wrongdoing, a CIA panel decided that they had acted in good faith, clearing them of any wrongdoing—despite a decision having been made by someone to search. Such a decision seems wrong on its face, the equivalent of having a defendant in a trial secretly recording the conversations of the prosecutors and the jury's informal discussions so as to be in a better position to counter any claim.

An appeal to one's failure to intent harm is not an automatic reprieve from doing what is morally wrong.

Intentions can make a difference to whether one is morally responsible, and we may thus be led to think that a person cannot be morally responsible for causing harm without intending to cause harm.

But that is a mistake. In our ordinary lives and, far more importantly for our purposes, in our professional lives, we can cause harm and be morally culpable without any intent to cause harm. Let us first look at the situation where there is intent to cause harm.

Moral responsibility because of intent

Suppose we discovered that a terrorist had meddled with the computer software on that plane, programming it so that at some time, like a hidden time bomb, some pilot would type in "R" for Cali and the plane would turn toward Bogota and thus directly into a mountain. We would have no trouble holding the terrorist morally responsible for the crash of the plane and the loss of 159 lives.

But that judgment depends upon at least three conditions being satisfied:

First, the person judged morally responsible must be capable of being morally responsible—sane, old enough, and so on. Toddlers are not responsible should they happen to cause harm, even serious harm. They are too young to know right from wrong and so too young to choose what is right over what is wrong. As any parent knows, the point at which a growing child can reasonably be held accountable is not marked on any growth chart. An infant can get away with murder, as it were, without being held morally responsible, and yet, at some point, we have no hesitation holding a child morally responsible.

The line has been crossed somewhere. Unfortunately, some seem never to cross that line. Sociopaths perhaps fall into that group.

And crossing the line does not necessarily mean that someone will stay across it. Reversion to infantile behavior is less rare than we would like it to be, and insanity and dementia and other forms of mental deterioration can preclude moral responsibility. Persons may plead an irresistible impulse, for example, when, due to some sort of mental impairment, they lose control of their actions. A kleptomaniac is capable of being moral and knows right from wrong, but cannot resist the impulse to shoplift. But we do not need to work out the subtleties of what is a complex matter—what a person needs to be to be morally capable—to understand that we cannot properly judge someone morally responsible if they are not capable of being morally responsible.

Second, the person must know that what he or she is about to do is wrong. Ambien is a recent addition to the remedies for sleeplessness, but some who have taken Ambien have driven to work while asleep, eaten in their sleep, and so on—only discovering what they have done after they have awakened and wondered where all the dirty dishes in the sink came from or gone to work in the morning and wondered who completed the report they were going to work on that day. Sleepwalkers are notorious for doing all sorts of things they do not remember and for which we cannot, reasonably or morally, hold them morally accountable.[31] Because they do not know what they are doing, they certainly cannot know that what they are doing is right or wrong.

As with the capacity for being morally responsible, there are complexities that a thorough analysis would uncover of what it is for someone to know that what they are about to do is wrong. Someone may know what they are doing, but not realize that what they are doing is wrong. We would then need to know whether they should have realized that and examine any reasons offered for why they could not have realized that. Someone may know what they are doing, but be incapable of realizing that what they are doing is wrong. Sociopaths may be an example of that as well as of the incapacity to be morally responsible.

We find such complexities as these in legal defenses. A person may plead that "… at the time of the committing of the act, the party accused was laboring under such a defect of reason, arising from a disease of the mind, as *not to know the nature and quality of the act he was doing*, or, if he did know it, that he did not know what he was doing was wrong."[32] The legal defenses mirror our moral defenses here. We would not hold someone morally responsible for an act under those conditions, but, as I say, there are complexities here that would need to be explored for a full analysis.

For instance, we may act to cause harm with less than full assurance that we will cause harm. We are not completely sure the bomb we fashioned will in fact detonate or that we will be able to penetrate a person's coat with the knife we intend to stab them with. We would still hold someone morally responsible for intending to cause harm—even knowing that they know that they cannot be fully sure they will cause harm. They are still doing what is wrong and they know that it is wrong.

But all we need for our purposes is to realize that in a clear case where someone does something and cannot know that what they are doing is wrong—the Ambien syndrome, we might call it—we do not hold that person morally responsible. They cannot intend to do anything wrong.

Third, the person must not only intend to cause harm, but also act so as to cause harm. A person may slip at the top of the stairs and, grabbing for support, knock someone down the stairs. We would not hold that person morally responsible for what happened to the person who fell if the slip were truly an accident. A body is sometimes an object—as when a person falls and strikes something else, a person, say—and we do not hold objects responsible for the harm they may cause. If a tree falls and hits a walker, we do not hold the tree morally accountable. It lacks the capacity for intent, and so, just as with people slipping through no fault of their own, there was no intent to cause harm and no deliberate act in order to cause harm.

What we would need to justify the claim that someone intentionally caused harm is thus that they are capable of being morally responsible, know that what they are doing will or is highly likely to cause harm, and do what they do in order to cause harm.

We can safely assume that these three conditions would be satisfied for someone who intentionally meddles with the plane's autopilot software to cause harm. That is why there is no room for moral doubt in judging that person morally at fault. If someone has the capacity to be moral, knows that the act in question is wrong, and intentionally does it anyway, we have all the reasons we need to hold the person morally responsible. We have a gold standard, as it were, for judging that someone has done wrong. Thus, had the software engineers deliberately introduced the software flaw, in order to cause harm, there would be no room for doubt in holding them morally at fault.

We need to make two distinctions here for clarity's sake. First, we need to distinguish between whether an act or omission is itself right or wrong, harmful or not, and whether its consequences are right or wrong, or harmful or not. A harmful act may have good consequences just as a good act may have harmful ones. Intentionally putting a flaw in the software is wrong, but its consequences would not be harmful if the flaw were discovered before it caused an accident. We would, however, blame a terrorist or the engineers if they intentionally introduced the flaw even if no harmful consequences occurred.

A second distinction we need to make concerns the character of the individual or individuals doing the wrong or harmful act. In describing someone as a terrorist, we already are making a character assessment: this is just the kind of person who would act to cause harm. But a software engineer? If a software engineer introduced a fatal flaw in the software, we would still have to ask, "Was this deliberate introduction of a software flaw expressive of their characters? That is, are they simply evil? Or is this a mistake, negligence, or what?"

When there is an unexplained airline disaster, one question investigators ask is whether something about the pilot or copilot could be relevant. Even if they had sterling characters up to the point of the crash, the question still needs to be asked and investigated. That is why investigators search the private computers of the pilot and copilot, looking for evidence that the crash was planned. And even if nothing is found, there is still the possibility that the crash was planned but nothing recorded about it or that a last-minute aberration triggered the crash.

So a full moral accounting of a situation requires that we determine if someone is capable of being moral and is likely to be, or not to be, because of their character, whether anything we know about them might suggest any aberrant behavior, whether they knew that what they were doing is wrong either in and of itself or because of its consequences, whether the person intended to cause harm and acted so as to cause harm.

If a person of good character intentionally does the right thing, knowing it is right, and it has good consequences, we have no foothold for moral criticism. If any one of these features is missing or less than sterling, we have at least a toehold for moral concern.[33]

Indeed, any variation in any one of them will affect our necessarily nuanced moral judgment. If the engineers deliberately introduced the software flaw, but the plane turned away from a sure accident rather than into one because of the flawed software, we would mitigate our overall moral judgment—though we would still hold the engineers morally accountable. If they introduced the software flaw by accident and tested the software thoroughly in the standard way, but failed somehow to find the flaw, we would soften our judgment about their moral culpability and try to determine what about the testing misled them into thinking the test was thorough.

If we have in mind the gold standard for moral responsibility— intentionally doing what is wrong or harmful—we may find it difficult to think of any situation where someone could be responsible without the relevant intent. We can readily imagine a person of bad character doing good—by accident or as a way of encouraging trust, for example. We can readily imagine a person of good character accidentally doing something wrong. We can readily imagine good acts producing bad consequences and bad acts producing good ones, but how, we may well ask, could someone be morally responsible for doing something wrong without intending to do what is wrong?

Moral responsibility without intent

In our ordinary lives, we can in fact readily find situations where we refuse to excuse someone who causes harm even when the person lacks any relevant intent. If I hit a child while driving too fast in a residential area, I am not off the hook morally if I say to a parent, "Oh, terribly sorry about that. I didn't mean to kill your kid." If I am engaged in target practice and a toddler wanders into my field of sight, I am not off the hook morally if I continue to shoot even if I do not intend to hit the toddler. I am morally culpable for putting the toddler at risk even if I think I am such a good shot that the risk is negligible or nonexistent. "I wasn't using him for target practice!" will not get me off the hook morally any more than "I was here first!" Even the best of shots can make a mistake, and continuing target practice while a toddler toddles close to one's line of sight is too risky to justify.

So it is not true that in order for someone to be morally at fault, the person needs to have had an intention to cause harm or needs to act so as to cause harm. In some very common sorts of cases, intention does not matter for moral fault and we do not need to act so as to cause harm. In the case of professional actions, or inaction, lack of intent to cause harm is generally even more clearly irrelevant—for at least two different kinds of reasons:

1. **Competence**—We read from time to time of an attorney who fell asleep during a trial, before the judge, the jury, and client. Before making a judgment, we do not ask, and do not need to ask, "Did the attorney intend to fall asleep?" One reason we do not ask is that that is such an unlikely possibility within

a courtroom, sitting on hard chairs, in front of a hard table while people are testifying and you are supposed to be paying attention. But we also do not ask the question because the attorney's intentions do not matter here. A sleeping attorney cannot hear the evidence or testimony so as to be able to rebut it or take advantage of it, cannot make objections to inappropriate remarks made by the opposing counsel, cannot, in short, properly defend a client. Attorneys are licensed, and a condition of their obtaining a license is that they have completed a course of work and passed an exam that at a minimum proves them qualified to practice law. A client has every right, in hiring an attorney, to expect at least a minimal level of competence, and that is not possible if the attorney is asleep during the client's trial. The lawyer's intention is irrelevant.

In a recent case, Texas state courts decided that, to quote the defendant's new attorney, "The state does not believe that you have a right to a lawyer who stays awake." In the case in question, the lawyer fell asleep a number of times during the trial, once for over ten minutes and once with his head on the table. The defendant's attorney said he was "as responsive in court as a potted plant," but the Texas prosecuting attorney had argued that there was no proof that the lawyer's sleeping made any difference in the trial.[34] When the defense attorney appealed to the U.S. Appeals Court, it said, "Unconscious counsel equates to no counsel at all."[35]

We might excuse an attorney in such a situation if the attorney had mistakenly taken an Excedrin PM instead of a regular Excedrin, or were suddenly overcome with narcolepsy, or was so exhausted from preparing for the trial as to be unable to stay awake during the trial. But these would be excuses, to be judged on their own merits as acceptable or not, and have nothing to do with whatever intentions the lawyer may or may not have had.

The problem for the attorney's client is that the lawyer is not at the trial just for show. A sleeping lawyer might as well be a dummy for all the good done for the client. A sleeping lawyer cannot even pull off impersonating a lawyer. We can readily imagine someone pretending to have the requisite knowledge and skills of a professional. *Catch Me If You Can* was a 2002 movie based on the life of Frank Abagnale Jr., a wonderful con artist with enormous skill in impersonating those in different professions—PanAm airline pilots, lawyers, physicians. But all Abagnale could do was impersonate: he lacked the capacities that are the essence of each profession. He could not fly a plane, create a will, check for a burst appendix. The failure to have the requisite knowledge and skills means that you are not a professional in the particular discipline in question, and so you can only impersonate a professional, not be one.

But even with a license to practice, and so, presumably, with the requisite knowledge and skills, you can fail to be competent. In fact, you can be so incompetent within your profession that we would rightly hesitate to honor you with the title that goes with being a professional.

At a VA hospital in Philadelphia, Dr. Gary D. Kao made mistake after mistake. In one case, he put most of the 40 radioactive seeds that were to kill a prostate cancer in a patient's healthy bladder. He corrected that mistake by rewriting his surgical plan "to match the number of seeds in the prostate" and then proceeded to implant more seeds in the patient, this time in the patient's rectum rather than his prostate. Over a six-year period, the VA hospital "botched 92 of 116 cancer treatments," with Dr. Kao apparently the attending physician in most if not all of these cases. It is difficult to call him "Doctor" Kao without irony. Someone who makes that kind of mistake, and makes it repeatedly, certainly lacks the specialized knowledge and skills necessary to be a physician even if they have somehow passed a qualifying exam and obtained a license to practice.[36]

We can distinguish between two different kinds of knowledge: knowledge that and knowledge how.[37] We can refer to a surgeon's knowledge that a kidney is not a gizzard, to a lawyer's that a continuance is not a dismissal and that a dismissal is not necessarily a dismissal with prejudice, or to an engineer's knowledge that a strut is a kind of brace or that the holding strength of bolts is different from the holding strength of welds. These are all examples of the kinds of information that a professional must learn to become a professional—and examples of knowledge that something is the case.

We refer to knowledge how when we reference the skills someone must learn. We learn to ride a bicycle. That is a skill. But in learning how to ride, we may have no knowledge about what the parts are called. We do not need that knowledge to learn how to ride. We learn how to use a software program that allows us to calculate stresses easily. We do not need to know anything else about the program to become skilled in using it.

So these two kinds of knowledge are distinct, but becoming an engineer, a surgeon, a lawyer, or any professional within a discipline requires both—both knowledge about the details of the specialized discipline and knowledge how to use that detailed knowledge to accomplish whatever ends the discipline is supposed to achieve. Surgeons do not build bridges. Engineers do not remove anyone's appendix. Doing either requires special knowledge and special skills—about the stresses steel or kidneys can face, for instance, and about how to cut carefully into a person's body or how to design a supporting strut on a bridge.

Becoming competent within a discipline thus means learning both information about the subject matter in question—the body, struts, the law—and the skills essential to using that knowledge. Even those who only minimally qualify for a profession must, we presume, have reached a relatively high level of competence: the training is long and arduous, the qualification tests fairly rigorous. We presume competence, that is, but that presumption can be rebutted, as it would be in the case of Dr. Kao at the VA hospital in Philadelphia. The point is that a presumption of competence is justified given the rigor of training and the licensing of universities. So when a surgeon cuts an artery thinking it a vein, or a lawyer fails to fill out a form properly so that a will is not legally valid, or an engineer fails to calculate a load properly, we hold the professional morally responsible even if the mistake was unintentional. The professional ought to know better.

Aristotle pointed out how difficult it can be to do the right thing— to act "at the right time, toward the right objects, toward the right people, for the right reason, and in the right manner."[38] The reason it can be so difficult is that for Aristotle being ethical requires learning a skill as well as knowing what is right and what is wrong, and learning a skill requires learning all manners of things, any one of which could go wrong.

Besides coming to have specialized knowledge and specialized skills, a surgeon, for instance, must learn to exercise enormous caution and care. Someone who fancies lightning moves, a thrust-and-parry cutting away of an appendix, for instance, or who has a lightning temper, moved to anger at the slightest problem, is not well suited to be a surgeon. Too much is at stake, and too much can go wrong, to risk irascible surgeons who fancy themselves fencers in an operating room. The process of making someone into a surgeon must weed out such traits—or individuals with such traits.

In the same way, engineers must learn to be risk averse, unwilling to resolve engineering problems in ways that risk unnecessary harm. They must be exceedingly cautious about the possibility of mistakes and so careful to check and double-check their calculations. Budding engineers who fancy lightning solutions or think themselves immune from making the errors that plague us all will not long survive in the rigorous training essential to making a competent engineer or in the real world of engineering should they somehow make it through that training. Aristotle says that "an expert in any field avoids excess and deficiency,"[39] and as he would put it of both such surgeons and engineers, "They are not acting in the right manner to be doing the right thing."

We can multiply such examples almost ad infinitum, there being so many ways in which we can fail to do the right thing in being an engineer, a surgeon, a lawyer, or any other professional. For instance, included in the requisite knowledge essential to any profession is the capacity to fill in the blanks, as it were, in your professional knowledge. A lawyer must "know the law," but that does not mean knowing everything of legal import—the substance of every case ever decided, for instance. It does mean knowing how to find out what is of legal import that is relevant to a case in hand. So one skill professionals must learn is how to learn what they do not know within their disciplines.

They must also have the capacity to keep up with professional practice. An engineer who relies wholly on what was available in four or five years of undergraduate work in engineering cannot claim competence for very long in the profession: new discoveries and techniques impact professional practice, and an engineer must keep up just like every other professional. Becoming an engineer by getting an engineering degree is not the end of a person's learning to be an engineer.

So having an intent to cause harm does not matter for a professional in at least those situations where the harm is caused by a failure to be even minimally competent. The ways in which someone can be incompetent are various and many—from not knowing what they are supposed to know as a professional to not knowing how to use that knowledge to not using it in the right manner, at the right time, for the right reason, and so on.

And, of course, the level of a professional's knowledge and skill is another matter that needs to be considered regarding intent. An engineer may not intend to cause harm, but is not competent enough regarding what is at issue—the strengths of various kinds of concrete, for instance—so as to avoid causing harm. We would hold such an engineer morally responsible even without any intent to cause harm. The situation for an engineer would be no different from that for a surgeon, say, who failed to see the difference between a kidney and a liver in situ or failed to realize that, as is the case in a small number of humans, the bodily organs are reversed, with the heart on the patient's right side, not the left, for instance. In both these cases, that of the surgeon and that of the engineer, whatever intent they may have had is not relevant if they fail to do what they ought to do as a surgeon and as an engineer.

So we hold professionals morally culpable when they lack the knowledge they were supposed to learn to become professionals. We hold them morally culpable when they have that knowledge, but fail to use it in the way they should have learned to use it as professionals. We hold them morally culpable if their level of competence falls below the level they really needed to do the job they were supposed to do and, as a result, they cause harm. We hold them morally culpable in all these cases even if their failures are unintentional.

So one ground for holding professionals morally responsible even for unintentional acts is that they are supposed competent. They do not impersonate professionals. They are professionals. A failure to live up to their professional responsibilities by being incompetent is a moral wrong whether they intend to cause harm or not.

2. **Special moral relations**—We have another reason for holding them morally responsible even without any intention to cause harm. We have eluded to this reason in some of the examples we have given. That is that professionals take on a special set of moral relations in regard to clients or patients or companies.

We read from time to time of a surgeon cutting off the wrong limb, a leg perhaps. Indeed, cutting off the wrong limb, or the wrong body part, seems to happen frequently enough that the error rate should give one pause if about to have such an operation. Surgeons cut off the wrong leg for an 86-year-old man in Lima, Peru and then had to cut off the other,[40] and a report about mistakes in 2005 says that surgeons in England removed the wrong disc in eight cases, amputated the wrong leg in five cases, took out the wrong hip in four cases, removed the wrong testicle in one case, gave a hysterectomy to a woman who did not need it, transplanted the wrong set of lungs into a patient, and circumcised a child who was not the patient needing circumcision.[41]

A surgeon's doing any of these things intending to cause the harm that resulted would be particularly egregious because the patient was presumably under an anesthetic and so helpless. More importantly, the patient was under the surgeon's care and so had every reason to expect that the surgeon would not take advantage of the situation to cause the patient harm, but would do everything possible to ensure that the patient was properly cared for. Intentionally amputating the wrong leg, even if neatly and carefully done, is hardly proper care.

Yet, even if the amputation were unintentional, we would still blame the surgeon. The person was the surgeon's patient. Surgeons take on special moral relations when they take on a patient. If a surgeon agreed to give me a needed operation, that surgeon would thereby take on a set of special moral relations. A drunk or hungover surgeon operating on me would be unprofessional—and unethical. Hacking at me and not cutting carefully, even at a spot where common surgical practice says to cut, would be equally unprofessional—and unethical. The surgeon asking someone else to perform the operation would be equally unprofessional—and unethical. I empowered that surgeon to operate on me and not any substitute. In putting myself in this surgeon's care, I am obligating the surgeon to operate on me at least to the minimal standards of the profession. If the surgeon amputated the wrong leg, even unintentionally, I would hold the surgeon morally responsible. "Whoops, sorry about that! I didn't mean to cut that one off!" will not work to get the surgeon off the moral hook. Minimal competence in surgery requires getting the problem right and doing right by the patient in solving the problem. Cutting off the wrong limb fails on both counts.

We can readily multiply such examples of various professionals failing those they work for. We hold professionals morally blameworthy, and rightly so, if through inattention, carelessness, neglect of advances in their field, or a failure to do whatever they are obligated as professionals to do, they cause gratuitous harm to those they have taken on in a professional relationship.

Engineers, for instance, come to have a professional relationship with employers, with clients, and with other engineers and other professionals with whom they are working to accomplish whatever it is they have contracted to do. They can fail to fulfill their parts in those relationships in a variety of ways, as we have seen, and when they fail, intentionally or not, we rightly hold them morally accountable.

We have presumably all experienced examples of free riders, those who are assigned to a project with us but fail to do their part. They are trying to take a free ride on the backs of those who are doing the work, trying to get the credit without the work. Any group working together is always open to the possibility of some member or members of the group shirking their duties. That may or may not be harmful to the project the group is engaged in. It may even be a relief to have someone who is not particularly helpful not trying to help. But even if we may quietly applaud their shirking, we still must hold them morally accountable. They owe it to the other members of the group to do their share, and even if they unintentionally fail their fellow engineers by not doing their share of the work assigned or not doing it in a timely way and not doing it competently, we would hold them morally accountable.

We have also presumably all experienced examples of where someone who is supposed to be part of a team fails in some way, intentionally or unintentionally, and so harms the project all are supposedly engaged in as well as putting extra burdens on their fellow professionals.

Some examples of such failures are bewildering and so bizarre as to be fit subjects for a Monty Python skit. In one example, a surgeon and an anesthesiologist ended up wrestling on the floor of the operating room while their elderly patient was under a general anesthetic. They began to argue just before the surgeon was to start the operation.

One thing led to another, and they were soon on the floor while the nurse looked after the patient.[42]

We would hold them morally accountable both for being incompetent and for failing each other. They are supposed to work together as a team, and when they fail, they not only fail their patient, they fail each other and they fail the test of competent medical practice. We have, as it were, a double whammy of a moral fault, two different grounds for holding someone morally accountable even without any intent on their part to cause harm. For two different reasons, therefore, we can hold professionals morally responsible for what they do, even without any intent to cause harm. We presume competence on the part of those who have qualified to be professionals in a discipline, and when the harm they introduce is caused by their incompetence, we rightly hold them morally accountable. We also hold them accountable because of the moral relations they have taken on in having a patient or a client or an employer or fellow professionals with whom they are supposed to work.

Those software engineers

No one suggests that the software engineers deliberately designed the autopilot software in that Colombia aircraft to mislead pilots, but that makes no difference to our holding the software engineers morally responsible for having created the software and for the subsequent crash and loss of 159 lives. They failed the company that hired them. They failed the pilots who relied on them. And they failed the passengers whom they put at such great risk by having designed software with a fatal flaw.

That flaw led to the crash, and the pilot was not responsible and did not hit the wrong key. As I suggested, we might hold the pilot responsible for not double checking to be sure that in hitting "R" he engaged the autopilot to land at Cali, but clearly the bulk of the responsibility lies with those who designed the software that required checking because it was flawed. The software engineers were responsible, and this is not because they intentionally designed the flaw into the software but because, having designed the software with that flaw, they failed to design the flaw out.

They could have failed to design the flaw out for two different possible reasons:

1. One possibility is that they did not know it was there. We can hardly fault them for not designing it out of the software if they did not know it was there. But they ought to have known it was there. They created the software and so were better positioned than anyone else to understand it and see that they had written one line that made the default for the autopilot the closest beacon and another that made Bogota the default for "R"—although it is certainly arguable that it is often those most deeply engaged in an enterprise who are least able to back off to see any problems with it.

2. Or it is possible that they knew that they had designed software with competing lines of code, but failed to think through the implications of having two different defaults. But if they knew there were two different defaults, they ought to have thought through the implications of a pilot's typing "R" and suddenly finding the plane heading towards Bogota instead of towards the airport the plane had been heading for.

In short, either they knew of the flaw or they did not, and either way they were responsible for the subsequent problem that pilot had and for the subsequent loss of life and of the airplane.

How could they have gotten themselves into such a position? The answer seems obvious. Engineers design artifacts that solve problems. The artifact can be a piece of software, a tractor hitch, a door handle, a car. The kinds and numbers of artifacts that engineers design seem as numerous and diverse as anything in nature. Engineers are to check what they have designed to see if it works the way it is supposed to work in order to be sure that they have not overlooked something. An artifact that does not solve the problem it was designed to solve fails the most crucial of tests engineers are obligated to conduct before letting the design out the door.

Had the engineers thoroughly checked out the software to see how it worked, they would have discovered that typing "R" tells the autopilot to pick out Bogota and that the choice overrides the norm that the autopilot pick out the closest airport. It is difficult to imagine any engineer so incompetent as not to see that overriding the norm could cause a problem. One criterion for successful software for an autopilot is that it take over the controls of an airplane to land it safely in the airport the plane is supposed to land in. Checking out the software they created would have told the software engineers that their software failed to do that when "R" is typed—unless the plane was to land at Bogota. They did not need to know anything else to know that they had failed to design their software properly.

Had they tested the software, the engineers would not have needed to think through the implications of having two different defaults. They would have been able to see that typing "R" tells the autopilot to fly the plane to Bogota no matter where it is. That is one point of testing. It allows you to see things that otherwise you might miss.

In failing to realize they had introduced competing lines of code or in failing to think through the implications of having different defaults, the engineers are in good company. We often miss the obvious in our ordinary lives. We fail to realize we are locking our keys in the car until after we lock the door, for instance. We fail to think through the implications of what we are doing, not realizing until after the fact that something that may have seemed obvious at the time turns out to cause problems elsewhere.

The software engineers are also in good company professionally. Microsoft has become famous in part for releasing software with all sorts of bugs. The more complex something is, the more likely it is to have flaws and the more unlikely it is that anyone will notice the flaws—particularly when the flaws are not in individual parts, lines of code in this case, but in the combination of distinct parts. Our modern technological lives are filled with flawed artifacts—rearview mirrors that drop off the windshield because the glue fails, two-ton concrete slabs that fall off tunnel ceilings, cell phones that cannot be held without risking pushing buttons on their sides that will interrupt calls, remote controls so complicated that we struggle to find the mute.

But, again, that is the point of testing. Had the software engineers tested the software as they should have, they would have found the problem, and, having found it, they should have fixed it. There is no reason why typing "R" ought to override the normal default setting and direct the autopilot to fly the plane to Bogota. That line of code is not essential to the autopilot software, and we can readily predict an accident given the two different defaults. We know that because of the flaw in the software, some pilot somewhere at some time would type in "R" for the beacon at one airport and the plane would turn toward Bogota. That creates an emergency for the pilot and could well lead to a crash—as, indeed, it did.

So we quite properly hold the software engineers morally culpable for failing to test the software, discovering the problem, and then redesigning the software to remove it. We can now see that it makes no difference whether they unintentionally introduced a flaw into the autopilot software. They failed the competence test, failing to reach even a low level of competence in designing that software, and in failing to produce an artifact that worked to solve the problem it was supposed to be designed to do, they failed to fulfill the responsibility they had to those who hired them—as well as to the pilots who had to use the software and to the passengers whose lives depended upon its working properly. Intentions can certainly matter in making moral judgments, but unintentionally causing harm is not always a moral excuse—particularly for professionals.

External moral relations

For two very different reasons, then, we hold the software engineers morally responsible for the accident. First, they are software engineers, and there are moral responsibilities they have that are internal to their profession just as there are moral features internal to every profession. Software engineers are supposed to be qualified to create software that solves whatever problem it was they were hired to solve. They also failed to exercise the special care professionals are required to exercise. Not testing the software is as irresponsible as testing it and not seeing the problem it would create for a pilot. The surgeon who cuts off the wrong leg, the dentist who, daydreaming, drills through and ruins a tooth, the engineer who designs software with a fatal flaw—these are all professional failures, examples of incompetent professional practice.

The engineers' failure of competence means that their design solution was a failure as well. The engineers are responsible, as we saw, because in becoming engineers, they took on two different sets of moral relations:

1. **Role morality**—In becoming an engineer, a person takes on a set of role-specific relations having to do with the practice of engineering—for example, ensuring that calculations are made correctly. The most obvious examples of what taking on that role requires is to be found in the proper use of the proper rules of skill for the engineering profession.

2. **Internal**—The intellectual core of engineering is solving design problems, and because at a minimum, ethically, an engineer ought to cause no unnecessary harm, solving design problems requires ethical considerations if only to avoid a solution which causes unnecessary harm.

In regard to the software design for the autopilot at least, they failed to reach the level of competence necessary to practice engineering properly, and so they failed to solve the design problem they were given.

They also failed, morally, in a third way. Those software engineers were either hired to solve that problem or given that problem by their employer. They were given the job because they were supposedly qualified to do it, and having created the software, they were, as I have pointed out, the best positioned of anyone to realize that the software was flawed. If they did not catch it, no one else was likely to catch it. In taking on that job they entered into a special set of moral relations with those responsible for providing the autopilot, and in creating flawed software, they failed to fulfill their moral obligation to the company for which they designed the software.

We thus need to add a third set of moral relations:

3. **External**—When engineers hold an additional position— wearing another hat as an employee, contractor, or manager, say—they take on the additional ethical relations required in those additional roles. So they also can have additional kinds of ethical problems.

The failure to do what they were hired to do or what their employer had assigned them to do is an external moral failure because it is a failure they have as contractors or employees who happen to be engineers. It is not internal, not one they have just because they are engineers. The failure to check the software properly or, if they did, to find the problem or, if they did, to correct the fault is an internal moral failure. It is part of what it is to be an engineer, of the essence of the profession, that engineers check their design solutions to ensure that they in fact solve the design problem. But the failure to do the work properly that they were hired to do is not a failure the engineers had as engineers. It is not part of their profession to be an employee, a contractor, or a manager. They may not be able to make a living without taking on one or more of those roles, but even in penury, they will continue to be engineers.

It may help to contrast the role morality of engineers here with that of, say, servants. It is part of what it is to be a servant that one serve others. One can be trained as a servant without becoming a servant, but to be a servant, you must be employed as a servant. Being an employee of a certain type is part of what it is to take on that role. And servants are not unique in this regard. A person who is elected to Congress to represent a district is a representative, and the role-morality of that position requires that they represent that distinct. Whether they do or not, and whether they do so well or not, are different issues. But the role itself is like that of a servant: it requires a relation with others. Engineering does not, and so in becoming an employee, for instance, an engineer takes on additional moral relations I consider external to what it is to be an engineer.

I have been arguing that ethical considerations are internal to engineering, and so it is of some importance to distinguish between the kinds of moral problems that arise for an engineer as an engineer and those that arise for an engineer as an employee. An engineer can fail to do a competent engineering job as an employee, but then the failure is subject to moral criticism of two completely different moral grounds, that the engineer failed the test of competence as an engineer and that the engineer failed the test of competence as an employee.

That the same failure can be criticized on two different kinds of moral grounds should be no cause for surprise. If I hire someone to mow my lawn, and they do a terrible job of it, they have failed me in two ways. First, they did not do what I hired them to do, namely, mow my lawn according to at least the normal standard (e.g., do not miss swaths of grass, do not leave stalks where the grass has been cut), and, second, they did not do what we should properly expect any competent mower to do (e.g., not miss swaths of grass; and so on). We expect mowers, dentists, surgeons, or engineers to be able to do what they represent themselves as being able to do. If we hire them, we expect them to do what they are supposedly competent to do. Being competent is one thing, doing what they were hired to do is another, and a professional can fail in either way or in both.

The software engineers had clients whom they failed, and they had software they failed to test thoroughly. The former failure is clearly a moral as well as legal failure. They represented themselves as able to produce a design solution that worked, and they failed to do that. That failure, however, does not call their professional credentials into question. It is not their failing their clients that puts their credentials at risk, but how they failed their clients—by failing to test thoroughly the software they designed or, having tested it, by failing to see that it was flawed. They should have tested the software or, testing it, seen the flaw not just because they were obligated to those who hired them, but because they are software engineers. Their failure *as* engineers is also a moral failure, and it does put their professional credentials into question. We must wonder how an engineer who designed software of such importance for the safety of so many could have failed to test it thoroughly or failed to see the flaw. We must ask, "What kind of engineer is that?" The implicit implication of the question is that that is no kind of engineer at all. That implication can readily be rebutted by showing that the mistake in the case in question was an aberration, not at all typical of the kind of work the engineer generally does, but given the kind of mistake made, it is an appropriate question to ask.

How ethics enters engineering

So how does ethics enter into engineering? It certainly enters when an engineer takes on those moral relations that come from working for a client, being an employee, taking on a contract, or even from working with other engineers as part of a team. When engineers work for a client, they are obligated to represent the engineering problem clearly, ensure that the problem identified is solved without creating any new problems, and so on. They take on special obligations when they work as a member of an engineering team. They are obligated to resolve disputes between themselves in an amicable manner, for instance. They are obligated to do their part and not free ride on the work of others. When they fail to fulfill those obligations, their failure is a moral failure.

Yet examining the ways in which ethics enters engineering through taking on those moral relations will only tell us what kinds of moral relations a person can enter into *as something other than an engineer*— as an employee, as a member of a team, as a contractor. All sorts of moral issues arise when a professional takes on such moral relations, but examining those issues will not answer the question, "Does ethics enter into engineering itself?"

What must engineers do whether they work for a corporation, work with other engineers as part of a team, or work as independent contractors? What is it that makes them engineers? What is the reason that they—rather than, say, mathematicians—were hired as engineers or are working, as engineers, on an engineering team? Does what they do as engineers require moral judgment?

Making design decisions is at the core of engineering. That is what engineers do: they solve design problems of a certain sort. In doing that, they make decisions—about what lines of code to use, about how to arrange the burners on a stovetop, about what materials to use, and so on. In choosing one design over another, they are choosing, as we saw, one set of values over another, and some of those values are moral. This is what we may call the argument from design in claiming that ethical considerations are internal to the profession of engineering.

As I also argued, these decisions will have their effects once they are realized in an artifact. Some of these effects will be good, some will not, and engineers are responsible—*at a moral minimum*—for making design decisions that do not cause unnecessary harms. We may call this the argument from effects.

So in two ways the solving of a design problem requires ethical judgment, but ethical considerations are also internal to engineering because engineers can fail to use properly the proper rules of skill that they must use as engineers.

So the software engineers did not make a morally neutral set of decisions when they created software with two different defaults and then failed to find the flaw. They were morally wrong *as engineers*. They also were morally at fault as employees with a responsibility they took on as employees to do what they were assigned to do properly. The harms that came from the autopilot software were unnecessary. Designing the software correctly to begin with, or testing it to find the problem and then correcting it, would have precluded the plane's turning from its flight path and so prevented the crash and the death of 159 people.

It does not matter if their failure was intentional or not. A failure to fulfill a professional obligation is a moral failure whether intentional or not. A failure to do what engineers are supposed to do as engineers is a moral failure, whether intentional or not, and a failure to do what an employee or contractor is obligated to do is a moral failure as well, whether it is intentional or not. Intentions do not matter morally in regard to the inner morality of engineering or, in some cases at least, in regard to the external morality of a professional.

An evil genius of an engineer

Engineers could, if they wished, readily cause enormous harm. They are at the center of our technological lives, designing everything that we might think of as exemplars of those lives—iPads, our highway system, mobile phones, stoves, trains, planes, cars, our electrical grid, and on and on. The list is almost endless, there being few if any artifacts of our lives that have not been designed by engineers. It is, indeed, difficult to imagine any artifact that has not been touched by an engineer.

To see how much we owe to engineers, to engineers going about their jobs with the competence they expect of themselves and of each other, we can suppose an evil genius of an engineer who intentionally takes advantage of decision points in solving design problems to introduce harmful results. In making this supposition, we are putting to one side any havoc an evil genius of an engineer could cause by failing to fulfill the moral relations taken on as an employee, a colleague on a team, a contractor. The havoc such an evil genius of an engineer could cause in such cases would be no different than the havoc caused by an evil genius of a manager, or an evil genius of a worker putting together whatever an engineer might have designed. We are instead concentrating upon what a evil genius of an engineer could do as an engineer to cause the harm, and we are obviously supposing an intent to cause harm—as a way to highlight not why engineers should have an intent to do good, but why engineers do not need an intent to do good provided they are competent and, obviously, do not have an intent to cause harm.

We all, at some time or another, have experienced difficulties with some artifact or other where the fault was clearly ours. A really evil genius of an engineer would so design artifacts that even the brightest, most knowledgeable, and highly motivated would be provoked into making a mistake that would cause harm. When instantiated in artifacts, such designs would mislead the best as well as catch up all the rest of us who are not the brightest, most knowledgeable, and most highly motivated.

The most perverse evil genius of an engineer would deliberately design artifacts that seem to tell us to do one thing to get done what needs doing, but that cause harm when we do it. We are all familiar with such artifacts and have examined some at some length—stove top layouts that falsely signal us to use one knob rather than another, doors that appear to open one way but open another, door handles that appear to turn one way to open, but turn another, and so on. None of these particular examples may seem to cause great harm, but each has the potential to do so, and, as the flawed software in the autopilot of that Colombia airliner makes clear, it is easy to find examples of what a perverse evil genius of an engineer could do that would cause great harm.

We shall examine more examples as we go on, but the point, I hope, is made. An evil genius of an engineer could do much harm in this world of ours because there are many ways, in solving a design problem, to introduce features that produce harm. The decisions engineers make in solving design problems are not morally neutral, that is. Those decisions instantiate moral judgments—whether engineers are aware of that or not.

An evil genius of an engineer who intentionally produced such designs should be drummed out of the profession, but whether flawed decisions are made consciously or not does not matter to our judging that they are morally wrong. We often make moral decisions without thinking about them and without having the time to form an intent to do or refrain from doing anything. When someone beside me stumbles, I reach out to help—without thinking about whether I should or should not, about what the consequences of my helping or not helping would be, about how I must look to others as I serve as a role model or not, and so on. Just so with professionals. Whether they think about what they do or not, they are still morally liable for what they do, liable to be blamed and praised, that is—especially when others are relying on their special expertise, their special knowledge and skills, to solve competently whatever problem they have been given.

We seem to notice their moral accountability only when something goes badly wrong, and we ask, "How could someone who claims to be an engineer have done something like that?" But engineers act morally whenever they solve a design problem, whether harm is produced or not. They either make the right decision, the wrong decision, the best of a bad lot, the worst of a bad lot. They choose a solution, but have no choice in the matter whether they act morally.

Their choice will reflect a particular configuration of values, and they are making a moral decision in choosing that configuration over others. It is not a morally neutral decision to emphasize cost over efficiency, for instance, or ease of manufacture over safety—if only because differing design choices reflect different configurations of values, with different implications for which values are being favored and which frustrated. Different design solutions embody different sets of harms and goods. Those different solutions, when instantiated into artifacts, are going to have different sets of effects, and engineers have an obligation, at a minimum, to minimize the harmful effects— to repeat the point, yet again.

6

Permitting, encouraging, and provoking errors

It is no surprise that engineers generally choose designs solutions that do not cause unnecessary harms, but if so, what is there to gain from pointing out that in doing that, they are making ethical judgments? It looks as though all I am doing is asking engineers to look at what they are doing in a different way that will not in fact change at all what they are doing. But that reading would be a mistake.

The gains of seeing design solutions as ethical choices permeate engineering practice, from our understanding of what counts as a design problem to what counts as a harm to what counts as a design solution. We will look in this chapter at three examples—designs for a trunk lid, an X-ray machine, and a defibrillator—where the design problem was not fully articulated. In these cases, the engineers failed to think through how these artifacts were to be used. We shall examine what count as harms and design solutions in the next chapter.

The argument so far

We now know that:

- Design problems are open-ended: the problem does not itself determine the solution.

- Design problems thus leave conceptual space for creative imagination.

- The chosen solution, whether creative or not, will select one set of values over another (the argument from design) and, when realized in an artifact, produce one set of effects over another (the argument from effects).

- The choice of a design solution is thus an ethical choice, the set of values chosen reflecting one of many and the effects producing more or less harm than other choices.

- So engineers make moral judgments in solving design problems—if only the judgment, made self-consciously or not, not to solve the problem in a way that will provoke errors on the part of those who use the artifact that embodies the design solution.

Because engineers have professional responsibilities as engineers, just as, say, surgeons have professional responsibilities as surgeons, we also know that:

- Engineers are morally responsible for the judgments they make in solving design problems—whether they intend to cause harm or not.

In engaging in the core intellectual enterprise of their discipline, solving design problems of a certain sort, engineers are thereby making ethical decisions.

We also know that:

- To become an engineer, a person must learn the rules of skill essential to being an engineer.

- To be an engineer, a person must use the rules properly.

- A failure to use the rules properly raises an ethical red flag: the role-morality of an engineer requires a competent use of the rules of skill essential to the profession.

The best examples of how ethical considerations can enter into the intellectual core of engineering are error-provocative solutions. The argument from those examples is simple: if an engineer could intentionally choose error-provocative solutions in solving design problems, they would surely then be making a moral judgment—not a good one, of course, but a moral judgment nonetheless. So in choosing solutions which are not error-provocative, they are also thereby making moral judgments, at the least judgments not to cause unnecessary harm. These are not trivial judgments.

Suppose that all the world's engineers were evil geniuses, striving to introduce as much harm into our technological world as they could. They would have no trouble wrecking havoc in our technological world. They could create a world where nothing worked—no phones, no cars, no stoves, no furnaces, no water heaters, nothing at all. They could create a world in which everything looks as though it works, but fails—cars start but then stop, stoves turn off as soon as we turn them on, phones ring but cut off when answered. They could create a world in which everything appears to work one way, but instead works another way—where everything was error-provocative, producing harm whenever any operator did what the design of the artifact signaled ought to be done.

If they were really perverse, they would create a world where enough things worked the way it appeared they worked and worked well enough to give us sufficient confidence to move about and try to do things, not always wondering whether something will work this time or whether that new artifact in our lives will cause the sort of frustrations we hope to avoid. We would have enough regularly in the way things work not to find ourselves always worried, but we would then find ourselves caught up short when things went wrong— when nail guns would randomly misfire, sending nails into us; when steering wheels would come off in our hands without warning; when gas furnaces would leak in completely unexpected ways, with subsequent explosions; when airplanes would crash because some random part or other would fail unpredictably.

Engineering so permeates our lives that it is difficult to imagine what the world would be like if engineers tried to cause harm. Even a few well-placed evil geniuses, choosing just the right weak points in our technological world—the electrical grid, computer software— could turn our world into a chaos of malfunctions. We have a glimmer of what kinds of harms a few well-placed and very adept individuals can cause by seeing what can and has gone wrong with the internet, especially with what was supposed to be secure information.

We do not live in such a world in which we have a chaos of malfunctions because—no surprise here—engineers generally do what they ought to do. They make choices in solving design problems that minimize harm—such as putting the knobs for the right-hand burners of a stove top on the right-hand side and those for the left-hand burners on the left-hand side. In order to be moral, as we have said, they need not have thought through whether those choices minimize harm or encourage or provoke it. Their choices will have their effects whether they think about them or not, and, rather obviously—because we do not live in a world of artifacts designed by evil geniuses—they generally make the right choices.

So what difference does it make?

Pointing out that ethics is internal to engineering practice makes explicit, I have said, what is implicit. We are only looking in a different way at what engineers already generally do. As I have just pointed out, engineers generally do just what they ought to do, but if so, one may well ask, what difference does it make to engineering practice to think about solving design problems morally? If engineers generally do the right thing without explicitly thinking about ethics, why should they think about ethics? What is the gain?

Like ethics, the gains permeate engineering practice.

1. **What counts as a design problem?** If engineers look at their solutions to design problems as making moral choices, they can see problems they might not otherwise see. Of all the things that can go wrong to cause an accident, the worst for an engineer is for an engineering artifact—software, for instance—to be implicated because it is error- provocative.

To point out the obvious, being wholly at fault for causing
what can be great harm—159 dead, for instance—is not a
good position for anyone to be in, but to avoid that kind of
problem, engineers need to think about how an artifact will
be used and query various design solutions to determine if
they are easier or harder for those who will use them when
instantiated in an artifact. A design problem is not just a set
of specifications to produce a certain end, but, once solved
and instantiated in an artifact, will have effects in the world.
Engineers are responsible, at a minimum morally, for ensuing
that their designs do not themselves provoke unnecessary
harms when realized in an artifact.

To repeat, it is unnecessary harms that are to be avoided. It is difficult
to imagine a design solution that avoids all harms. No matter what
the material from which it is made, there will be a carbon footprint in
obtaining it, another in manufacturing the artifact or the artifact in which
it is to be placed (as software is placed in computers), and so on. But to
avoid unnecessary harms, we need to expand our view of the nature of a
design problem and consider the whole range of harms associated with
any engineering solution. They are all open to moral consideration.

2. **What counts as a harm?** We have focused on simple
   design flaws that produce harm for those who use the
   artifact. Error-provocative designs are the most striking.
   When even the most intelligent, most highly trained, and
   most motivated of operators are provoked into making
   mistakes because of the design of the artifact in question,
   the design is at fault and the engineer who chose that design

is responsible. By focusing on such an example, however, we limit our understanding of the kinds of harms engineers need to avoid. Concentrating upon error-provocative designs can itself provoke an error on our part in understanding all the ways in which engineers can introduce harms into the world. Many engineering artifacts, perhaps most, are not properly described as artifacts that can mislead an operator—bridges, for instance.

The design process is not limited, that is, to determining how something should be designed to be used properly. Engineers also need to consider what materials to use (and so how dangerous it may be to get it and how much harm is involved in obtaining it), how complicated it will be to make the artifact (and so how much energy and time and money will be consumed), how complicated and costly it will be to store the artifact until it is sold, how long its useful life is, how easy or hard it is to repair and at what cost, what will happen to the artifact once its useful life is over (and so how much of the artifact can be recycled and how easily), and on and on. These are choices engineers ought to make in solving a design problem, and none is ethically neutral. The engineer who designed mercury switches presumably did not think through what would happen when the switch broke and spilled mercury or ceased working and needed to be discarded.

3. **What counts as a solution?** We have focused on what engineers ought to do *at a minimum*. "Do no unnecessary harm!" is the bottom moral line, and it is at the bottom. No one wants someone only minimally competent—whether

an engineer, a lawyer, or a surgeon. We may presume that professionals can be ranked on a bell curve of competence, going from the most brilliant to the good to the worst. Our educational requirements are, or ought to be, such that the worst professional is still pretty good, competent enough, if an engineer, not to design error-provocative artifacts. If not, we ought to move the floor higher.

Neither society nor any engineer ought to be satisfied, however, with solutions to design problems that display only minimal competence. We ought to presume that anything can be made better—design solutions and us as well, and though we cannot obligate anyone to do or be the very best, we should hope that every engineer would look on each solution as less than optimal, even if they cannot see how to make it better, and look upon every stage of their lives as a stage for improvement as well.

We can think of error-provocative designs as being at one end of a spectrum of possible design solutions with foolproof designs at the other end. A foolproof design is one that even the most unintelligent, untrained, and unmotivated cannot screw up—at least for artifacts that require operators. Just as engineers ought to avoid error-provocative designs, they ought to strive for foolproof designs. That would be to aim for the very best—at least in terms of an operator using the artifact. Unfortunately, as we all know all too well, there are too many different kinds of fools to design something that is proof against all mistakes. We all have to shake our heads sometimes when we hear of some mistake someone has made that, we would have thought, no one could possibly have made. "What were they thinking?" is a rhetorical question in such situations because we have no idea what they could have been thinking—or even if they were thinking. So a foolproof design is at best an ideal, but it is an ideal worth striving for. "Do no unnecessary harm!" should be complemented by "Strive for the best!"

In adding that imperative, we are adding to the set of moral relations an engineer may enter into. We have argued that the following relations are internal to engineering:

1. **Role morality**—In becoming an engineer, a person takes on a set of role-specific relations having to do with the practice of engineering—for example, ensuring that calculations are made correctly.

2. **Design solutions**—The intellectual core of engineering is solving design problems, and because at a minimum, ethically, an engineer ought to cause no unnecessary harm, solving design problems requires ethical considerations if only to avoid a solution which causes unnecessary harm.

These are internal because they are relations an engineer is in by virtue of being an engineer. We then noted another set of relations an engineer may or may not be in.

3. **External**—When engineers hold an additional position— wearing another hat as an employee, contractor, or manager—they may have ethical problems in that additional position or, more germane for our concerns, between what the positions require and what they are obligated to do as engineers.

In saying that "Strive for the best!" should complement "Do no unnecessary harm!", we are adding a fourth that an engineer may or may not have:

4. **Aspirational**—Engineers should always strive to better themselves as engineers, improving on their past design solutions, learning from their mistakes and the mistakes of others how to avoid errors, keeping up with the latest engineering techniques, understanding how new materials can make for better solutions, being dissatisfied, that is, with being merely competent.

Someone can be an engineer with no aspirations at all other than to be anything but mediocre. No engineer need to strive to be better to remain an engineer however much we may be saddened to see someone talented enough to be an engineer choose not to continue to strive to be better. At some point, perhaps, should engineering practice change rapidly, someone who fails to keep up will cease to be someone we would want to hire as an engineer and we may even cease to call that person an engineer.

However unlikely such a possibility may seem for engineering, we find this happening in other professions with some regularity. New technologies and discoveries can fundamentally alter the trajectory of an entire discipline. Biology departments used to be dominated by field biologists who spent their lives hunting down new varieties or re-examining known ones and making sure they were classified correctly. The discovery of DNA altered biology in such a way as to make such classification almost a quaint byway for the profession— nice to do given the history of classifications, but unnecessary given how powerful a tool DNA is to identify and classify plants. We find the same sort of change in paleoanthropology where our capacity to date ancient bones has altered our understanding of our evolutionary history. So the idea is not far-fetched at least that significant changes in technology, for instance, may effectively phase out some engineers and an understanding of what engineers do. I suspect, but do not know, that few engineers now use a slide rule just as learning to use one is no longer on any course list for engineering students.

However that may be, we are putting to one side for our purposes everything except what is internal to engineering—role-morality and engineering's intellectual core, solving design problems.

We shall first consider what counts as a design problem and then, in the next chapter, the other two aspects of design problems we have articulated—what harms can be introduced and ought to be avoided if possible, and what counts as a successful solution. The division is to some degree artificial. If we fail to understand a design problem fully, we are far more likely to introduce harms we could have avoided and will fail to solve the problem successfully. So each example we examine could well have been placed under a different heading.

What counts as a design problem?

Aristotle said of being ethical that "going wrong is easy, and going right difficult; it is easy to miss the bull's eye and difficult to hit it."[43] He could just as well have said this of solving design problems. Of course, a design problem usually has more than one solution. So there is more than one way for an engineer to hit a bull's eye, but there are so many

variables that need to be taken into account that it is all too easy to
miss the target and fail to get things right.

We saw one way of failing when examining the software flaw
responsible for the crash that killed 159 people. The engineers failed to
think through how what they designed would work in practice. This
is not the only way in which engineers can fail to understand what
counts as a design problem, but it is a far more common problem than
it may seem. Each of the following examples illustrates that point.

1. **Cadillac trunk**—In some older Cadillacs, you are to lower
   the trunk lid to within a foot or so of the latch and then let go.
   A motor takes over and gently closes the lid. If you push the
   trunk down to latch it, the standard way of latching a trunk,
   you break the mechanism. Once the mechanism is broken, the
   trunk will not latch at all. So you end up riding around with
   the Cadillac's trunk tied down with a bungie cord or rope—
   hardly the upscale image Cadillac would like to convey.

Repair is costly because it requires taking out part of the trunk
compartment and the back seat to get to the mechanism. So you end
up with a cascading set of effects—the trunk latch broken, a trunk you
are unable to latch, and a costly repair, all because you or someone
else tried to latch the trunk the way we normally do.

The self-closing mechanism creates a problem waiting to happen.
We all know that sometime, someone, even with a warning not to
close the trunk by hand, will break the mechanism. The trunk opens
just the way normal trunks do, with no sign on or in it indicating it is
to be treated any differently than any other trunk. So someone fixing
a tire or putting in groceries will get no warning that the trunk should

not be closed the way normal trunks are. A single visit to a hotel with a doorman who takes your luggage and slams the trunk will suffice to do in your trunk and your wallet.

The problem with the trunk of these Cadillacs is not at all unusual. We have in part a user problem. Those who are most concerned that the trunk be closed properly, and best positioned to know how to close it properly because they can read the instruction manual that comes with the car, are not the only ones who will close the trunk, and even they will have to guard against letting old habits take charge. But there are others who will use the trunk—an auto mechanic getting a tire, that hotel doorman—and so there is a risk of harm.

We also have in part a legacy problem. People are used to trunks operating in a certain way. Change the way trunks operate, and some people are going to continue to try to operate them the old way, just by force of habit. No matter how many warnings the manufacturer puts in the instruction manual, or even on the trunk lid itself, someone is going to try to operate it the old way. We ran into this problem with that toaster lever that, when operated as we are used to operating levers on toasters, will break the mechanism.

So an engineer suggesting a new design needs to consider how things might go wrong because of past habits that will need to be changed. Engineering progress requires pushing the envelope of design and so forcing new habits upon us, but those old habits can cause significant harm.

In this case, the harm is primarily financial—the costs of the time lost, of the repair, of not having the car available. The engineers responsible for this trunk-closing mechanism failed to do anything to prevent those old habits from causing harm—no warning signs, no mechanism to prevent someone with those old habits from slamming it shut when it seems to catch for no good reason. There could have been a catch on the mechanism, for instance, that prevents someone from closing the trunk lid without touching a switch or lever, and that would be a warning built into the new design solution that would at least minimize the power of old habits.

Vehicles are a wonderful source of examples of error-provocative designs. "But," one may well wonder, "do these constitute moral problems?" The test is whether there is significant enough harm that could have been avoided, and the determination of harm is not limited to loss of life, for instance, but extends to any setback to an interest we have.[44] Our interest in regard to the Cadillac is to have a functioning car, without unnecessary expense or time spent without the car while it is being repaired. Closing the trunk lid as we normally do in Cadillacs with a self-closing mechanism produces a cascading set of harmful effects. Engineers should try to avoid all those harms if they can design such a mechanism without those attendant harms.

Because engineers should try to avoid all those harms, we need not get hung up on trying to find a line between morally significant and morally insignificant harms. It is a question that has stymied philosophers, and engineers do not need to get caught up in that query in order to identify the harms that would be produced by a design solution or consider alternative solutions that avoid those harms. What matters for engineers are only two questions: is there harm, and is it necessary? If there is harm that can be avoided, it ought to be avoided.

In the following example, there is no doubt that the harm ought to have been avoided.

2. **X-ray machine**—A large X-ray machine was built so that the patient lay on a table with the X-ray on an extremely heavy arm that extended over the table. The arm was as long as the table upon which the patient lay and wide enough to cover the width of the table. It could be rotated as well as raised and lowered so that the X-ray could be focused on a particular spot on a patient. At the end of the day, when the machine was shut down, the arm was automatically lowered to an inch or so above the table to keep the X-ray safe from harm when the cleaners came in and to keep the arm from being accidentally jostled.

X-ray technicians always go behind a lead shield so that they will not suffer the consequences of too many X-rays. In this case, the technician operated the X-ray from a console in a room completely separated from the machine. There was a door into the operator's room, but it was placed so that there would be no danger to the technician. There was, that is, no direct line of sight to the X-ray table or the arm.

The technician controlled the movements of the X-ray arm through knobs and switches on the console. Every movement was programmed by the software specially developed for this X-ray machine and the console.

One afternoon, after finishing up with the last patient of the day, the technician opened the door and called out to tell the man he could leave. The technician then shut down the machine. When the technician went out to leave work, the nurse asked where the patient was. "I told him he could leave." "Well, he didn't come through here."

They found him face down on the table, flattened by the heavy arm of the X-ray that had been lowered to just above the table.

In a separate room, the technician had not been in a position to see whether the patient had left, and nothing about the software required that the technician check to see if the X-ray table was clear before shutting down the machine and thus lowering the X-ray arm. The patient had not heard the technician, and, face down, he could not see the machine coming down to crush him as it was coming down to the table top.

A little thought about how such software would be used in practice would have revealed the problem.[45] The software was written so that the technician did not need to check on whether anyone was on the table before shutting the machine down. That was an accident waiting to happen—as it did. We know ahead-of-time, given such a situation, that someone is going to shut the machine down on a patient. What was needed was a check on shutting the machine down that required the technician to go into the X-ray room and hit a button or move a lever on the machine. That way the technician would have to check to see if the table was empty. Or a scale could be added to the table so that anything on it would be detected and that signal would prevent the machine from being shut down. We might find that the scale was faulty sometime, but that safety feature would prevent most accidents.

The general lesson for the software engineers who designed the program is clear enough. They needed to think through how the software would work in situ. "What," they needed to ask, "could go wrong here?" The most obvious things that could go wrong are that the X-ray could misfire somehow and burn patients and that the arm could be lowered onto a patient. So the software engineers ought to have designed the software to preclude, as much as possible, both potential harms.

The situation regarding the X-ray machine is the same as that regarding the autopilot software. The software engineers failed to think through what was likely to happen when the software was being used by those it was designed for—a pilot, an X-ray technician.

But the questions software engineers need to ask are not limited to what will happen when it is used by someone. They need to check the string of software, obviously. Free of faults? No mistakes? Works? But their questions are not limited to the string and any problems with it. They need to ask, for instance, how the software works with pre-existing software. This is a problem not unlike that physicians need to consider when prescribing medication that may not interact well with other medication the patient is taking. A recent example concerns Plavix, an anti-clotting drug given to those who have had a heart attack, and an anti-ulcer drug, Prilosec or Aciphex, generally given because Plavix can irritate the stomach. Those taking both drugs have a 25% higher risk of another heart attack. So software engineers need to consider whether new software will work properly with the pre-existing software into which it is to be placed. They need to ask, in addition, whether the software gives clear directions to those operating it. This is a problem like that of anyone trying to communicate information and avoid ambiguity.

There are so many variables that need to be examined in designing software that it is understandable how a software engineer may fail to think through those effects that are likely to occur when the software is put to use in practice.

Some may think it is difficult enough learning to think like an engineer, and here we are demanding that engineers put themselves in the shoes of those who use the artifacts they have created—to think like a pilot, or an X-ray technician. But that is not such a demanding challenge. They do not have to be pilots, only think through what it would be like to be faced with two different defaults when trying to land. They do not have to be X-ray technicians, only think through what it would be like to operate the X-ray machine with that software, including closing up the X-ray machine without having any way of ensuring that no one is on the X-ray table. To ensure that their design solutions do not cause unnecessary harm, engineers do not need to stretch their thoughts very far, as the next example illustrates.

3. **Defibrillator**—Joshua Oukrop was 21 when he died. He was on a biking trip with his girlfriend, called out from ahead, "Hold on. I need to …," and tumbled over backwards, dead, his defibrillator having failed to work when needed. He had a genetic heart disease and a defibrillator that was to "emit an electrical jolt to restore [normal] rhythm to a chaotically beating heart."[46] Mr. Oukrop's defibrillator shorted out.

The defibrillator failed because of the deterioration of the polyamide coating on electrical wires "in a component that sits atop the sealed part of a heart device. The component, called the header, is essentially a junction box connecting a unit's computer and power supply with cables, or leads, that carry electrical impulses to the heart." But "body fluids can slowly seep into the header, which is not hermetically sealed, and cause [the] polymide to deteriorate,...."[47] The deterioration means that the defibrillator will short out when it tries to send a life-saving jolt to restore the heart's normal rhythm.

The manufacturer, Guidant, discovered the flaw in 2002, three years before Mr. Oukrop died. It fixed the problem, but did not inform those who had had the flawed defibrillator implanted or their physicians. It announced the flaw only in 2005 when it discovered that the *New York Times* was publishing an article about it.

Guidant did not inform the patients with the defibrillators or their physicians because, it said, it judged "the risks, like infections, associated with surgical replacements outweighed the risks posed by the device."[48] Replacing the defibrillator, it claimed, was likely to cause more harm than leaving it in place—even though the harm of leaving it in place meant that some who relied on it to save their lives would die when it failed. It seems an odd juxtaposition, weighing the certain loss of lives against possible infections, but whether Guidant was right or not about its assessment of risks, that paternalistic response prevented patients from making their own judgments and precluded physicians from taking part in judgments about the health of their patients. Neither physicians nor patients gave informed consent that the flawed defibrillators not be replaced. Neither may have wanted to be faced with having to make such a choice, but it was theirs to make, not Guidant's.

Guidant may have made the decision not to inform the patients or physicians because it continued to sell the old model until, apparently, its inventory was gone. After the *New York Times* article, Guidant appointed an independent panel to investigate, and among its findings was the following:

> During a period of approximately one year after the corrective action was taken in response to the observation of arcing, more than 4,000 of the pre-mitigated devices continued to be implanted, approximately 1,300 of which were shipped from CRM's in-house inventory and the remainder in the possession of the sales force and in hospital inventories.[49]

As an attorney for someone suing the company might put it, with great sarcasm, mimicking their reasoning, "Replacing a flawed defibrillator with an equally flawed defibrillator is surely not worth the risk of an operation."

Guidant made at least two unconscionable decisions:

1. not to inform patients and their physicians of the flawed defibrillators that had already been implanted, and

2. to continue to sell the flawed defibrillators, knowing full well that they were flawed, knowing that physicians and patients could not know they were flawed, and knowing that the devices would put those new patients at risk of death when they failed.

It is more than a little disingenuous for Guidant to continue to sell the flawed defibrillator while claiming that it was riskier to replace the flawed device than to leave it in place. If the risk of replacing the flawed devices was greater than the risk of leaving it in place, surely the risk created by operating to implant a flawed device must be higher still since the risky operation creates a new risk for the patient because of the flaw in the device. So why would Guidant sell what it knew was a flawed device?

It is difficult not to think that Guidant was moved not to inform patients and their physicians because they wanted to sell the flawed defibrillators. That is, it did not inform the patients or physicians so they could make up their own minds about whether to replace the flawed device because if it had informed them, it would have had to inform them that the replacement devices were equally flawed. They would not likely sell any and would presumably have to write off 4,000 flawed devices at $25,000 apiece—a great deal of which was presumably profit since these devices are relatively simple and not that expensive to manufacture even if designed and produced correctly.

By the time Guidant announced the defect, two people were known to have died and over forty defibrillators had failed. Over 29,000 were at risk of their defibrillator failing just when it was needed. They thus faced that unfortunate choice: keep what is there and hope it works when it is needed, or have yet another operation to replace the defibrillator for another that may or may not work properly.[50]

Guidant's morally unconscionable behavior has had another effect, that is, the loss of trust that Guidant is concerned about the health of patients in need of a defibrillator and a subsequent wonder about the industry in general. Guidant was willing to write off the health of patients in need of a defibrillator in place of writing off its flawed devices, and if Guidant was willing to do that, what assurance do patients and physicians have regarding any defibrillator or, indeed, any other medical device?

The engineers who designed the defibrillator were equally at fault for failing to think through how it was to be used and failing to ask themselves a simple question, "Will the parts withstand implantation?" If you are designing something that is to be used in a hostile environment—clothing for use by those fighting fires, for instance—it is irresponsible not to test it in that environment to be sure that it can perform its task. Selling flawed defibrillators is as unconscionable as selling clothing for fire fighters that ignites upon contact with fire.

The possible harm from the flawed device is significant—death from a heart attack. It is particularly galling that the source of the harm is the very device that is supposed to save your life. What Guidant continued to sell was a false sense of security.

Without the details about any of the internal workings of the company, we cannot know exactly what kind of moral problem we have here. For all we know, this may be a situation where competent engineers wanted to test the device, but were prevented from doing so by management. That is not an implausible hypothesis given Guidant's moral climate. But whatever the details, we do know that the device should have been tested in situ.

This is an important lesson, one that needs to be emphasized because it has not been learned. A "new way of connecting defibrillators to the wires" had been developed, but it was not tested for how it will work in humans. The Food and Drug Administration (FDA) said that no testing was necessary because "the new wiring connectors are simply a design modification and not a new technology." The history of failures suggests otherwise.

It would perhaps be more accurate to say that the FDA approved testing the new method of connecting wires by waiting to see what happened after the defibrillators with new wiring were implanted in patients. This is an odd mode of testing a product, but it seems the preferred procedure and explains why so many drugs, for instance, are recalled several years after their introduction because they failed to work or caused significant harms.

We now know about one of Guidant's flawed defibrillators, but the problem was not an isolated incident. Two other models had similar problems of short-circuiting, and Guidant ended up recalling at least seven models.[51] Medtronic, another maker of defibrillators, introduced a new thin wire connector in 2004 that "began to fracture and fail at an unexpectedly high rate. By the time they were recalled, they had been implanted in some 235,000 patients," putting all at risk.[52] That number makes the 29,000 put at risk by Guidant seem like only a minor catastrophe.[53]

We should add that it is difficult to assess the risk to heart implant patients because we do not know how many have died because of a failure of their defibrillator. The number of deaths that we know occurred because of a failure with a defibrillator is probably significantly smaller than the number of deaths actually caused by failures. The defibrillators are mostly implanted in older people, and when they die, the cause is attributed to heart failure, and no autopsy would standardly be done to determine if the defibrillator had failed. So we do not know even roughly how many have died because of a flawed device. Without that knowledge, we have no way of assessing the risk of keeping a flawed defibrillator versus getting a new one. No one can answer the question, "What is the chance that my defibrillator will fail?" So Guidant was in no position to claim, as it did, that the risk of replacing the device outweighed the risk posed by the defective device. We simply do not know what the latter risk was, and the FDA is thus in no position to claim that because a new device is simply "a design modification and not a new technology," it is safe. The most it can say is that the new version is as safe as the previous models— which is not to say that it is safe.

The Cadillac trunk, the X-ray machine, and the defibrillator are examples in which something is wrong with the design solution the engineers adopted. It is not that the artifacts will not work. Anyone testing them will find that they work just fine. The trunk will close as it is supposed to close; the X-ray machine will close down to the table as it is supposed to; and the defibrillator will send the charge it is supposed to send when activated. That is part of the problem. When tested in isolation from the situations in which they will be used, these artifacts seem perfectly fine. Put into the situations in which they will be used, enough will fail to work as they were designed to work and so cause unnecessary harm, including death in the case of the defibrillator.

The problem is that the design problem for these artifacts was not fully articulated. The engineers needed to develop a defibrillator that could withstand implantation and still work, software that would require that the X-ray table be empty before it could close down, and a trunk mechanism that would not be so likely to break if the trunks are closed as we are all so used to closing trunks. Expand the description of a design problem to include how a solution would be used, and you can protect against such failures. Think here of that odd toothpick that was to fit on the end of one's tongue. It is difficult to imagine that solution being thought viable once we imagine anyone having such a sharp object on their tongue being dislodged and accidentally swallowed.

All these examples—the Cadillac trunk, the X-ray machine, the defibrillator—are examples of failures to think through how these artifacts will be used. These are all examples of how harms are caused to those who are to use the artifacts realizing the design solutions. At a moral minimum, as we have said, engineers are responsible for ensuring that their designs do not themselves cause unnecessary harm, and yet that is just what these artifacts do when they are put to use.

But in restricting ourselves to examples of how artifacts are to be used, we should not forget the other ways in which design solutions can cause harm. We have picked these examples because they are clear and most clearly make the point that solving design problems requires more than simply coming up with a grand solution, however creative. It requires at the last thinking through the potential harms realizing that solution may introduce, and those harms, as we shall see in the next chapter, are not limited to those created for those who use the artifact.

7

Harms and design solutions

As we saw, one gain in seeing design solutions as ethical choices is that engineers are encouraged to look downstream, as it were, to see what happens once a design solution is realized in an artifact. We concentrated in the preceding chapter on error-provocative designs, but examples of unprovoked harms are easy to find. Some of these occur because of the same failure to think through how an artifact will work in practice. Airbags are a prime example. But some occur because of a feature of designs that is not likely to be noted by any of us.

A design is a sign. Any design solution sends a signal, intended or not, about how it is to be used. It is a source of information that is sometimes so ambiguous that we cannot figure out how to use the artifact properly. Shower faucets are a standard example, yet another instance of a failure to look downstream at a particular way in which an artifact's design can mislead us.

But engineers also need to look upstream, for any design choice introduces the potential for harms through entailing another set of choices—what materials to use, how much can be recycled or must be lost, how easy it is to salvage what can be reused, and on and on, all issues of sustainability.

In addition, once an artifact is in place and we can assess more accurately its impact, upstream and down, than we could when it was just an idea, we have a moral obligation to change it to make it less harmful if we can. A design solution always carries with it a tentativeness. We strive, or ought to strive, for the best, and that carries with it the admonition, "Perhaps we could do better!" We shall consider road lines and signs as instances of how we can do better.

Unprovoked harms

We have been focusing on artifacts whose use provokes harm—the autopilot software, the X-ray machine that crushed the patient. There is no doubt about the nature, magnitude, or gravity of the harms—people died, an aircraft was lost—and no doubt about the engineers being responsible for that harm. They designed artifacts that would cause mistakes for even the most intelligent, well-trained, and highly motivated user, and so they are morally responsible for the harms their flawed designs caused.

Yet, as I have said, these examples themselves provoke too narrow an understanding of the harms engineers ought to cull from their design solutions. Not all harms are provoked. Some occur without any help from us at all. It was not any error on our part, for example, that caused many Fords to burst into flames. The fault was apparently with a seal in the cruise control switch that weakened over time and let in brake fluid, corroding the wires. The switch received power whether the vehicle was on or not. The corroding wires would overheat, igniting an electrical fire—even when the vehicle was parked. Letting off the brakes caused a small vacuum in the brake lines, and the vacuum caused the seal to invert, weakening over time.[54] It was not any error on our part that caused the treads on many Firestone tires[55] and, later, Chinese-made tires to separate, causing vehicular accidents and driver deaths.[56] It was no fault of ours that millions in the Northeast, in the United States, and Canada experienced a massive blackout in August 2003.[57]

Not all of these examples are obviously the result of engineering mistakes. The problem with the Chinese tires was that the manufacturer decided to leave out the gum strip that keeps the treads from separating. It is difficult to imagine that such a decision would be approved by any engineer knowledgeable about tires. It is much easier to imagine that such a decision is driven by the manufacturer's desire to cut costs and increase profits. That decision raises a question about what responsibilities engineers have when the companies they work for make decisions that engineers ought to find irresponsible. That is an external ethical issue, one that arises for an engineer as an employee, but it is an external ethical issue that has implications for the internal role-morality of engineers. How far into the manufacturing stage does the responsibility of an engineer extend beyond providing a design solution to a problem?

In any event, whatever the source of the problems, the examples illustrate well how engineering permeates our lives and why it does not take an error-provocative design to raise an ethical issue for engineers. The range of possible harms, the gravity of those harms, the kinds of harms, the numbers of those affected by these various harms—all are missed if we consider only those design solutions which provoke errors on the part of those who use the artifacts that embody those solutions.

We will consider in this section two examples of harms, both drawn from ordinary objects, that require no mistakes on our part to harm us or put us at greater risk than necessary. The examples are both drawn from that rich source of problematic design solutions, the vehicles we drive.

1. **The clutch pedal**—In the Mazda RX-8, the interior edge of the side wall to the left of the clutch pedal curves in to form a lip at just the height of the pedal top. The lip is so close to the top of the pedal that when you try to push the clutch petal down to shift, your shoe gets hung up on the lip unless you are very careful not to push the clutch down in the center, but on its right edge—a maneuver which itself causes problems. If you fail to do that, your foot stops before it gets started. The lip is a barrier to pushing in the clutch. If, as you start to shift, you move your foot farther over toward the right to avoid the lip, you can end up putting on the brake when you try to shift. Only if you are lucky enough to have very narrow feet can you shift gears in this Mazda without a serious difficulty.

It is not as though it is the user's fault that the clutch pedal causes problems. Some people have wide feet, some have narrow feet, and it would never occur to us to fault the one or the other for the widths of their feet. It is just a fact. There is no doubt an average width, but it appears that even someone with a foot of average width would have trouble using this clutch pedal, and it is not as though a user missed a sign. This Mazda example is like the Cadillac trunk in that neither sends a misleading signal to the driver or user. Neither sends any signal at all. So we have no warning of anything problematic.

This problem of a problematic artifact that catches us by surprise is, unfortunately, not uncommon. I would suspect that mechanics could easily supply a raft of examples from their work on automobiles—a bolt that unscrews clockwise rather than counterclockwise, to repeat an earlier example. Here is another one.

2. **Airbags**—When engineers strive for foolproof design solutions, they are not just trying to stymie fools. They are striving to produce a solution which will ensure that no operators are harmed by the artifact. The first-generation airbags are an unfortunate example of how engineers, in trying to make drivers and passengers safer should an accident occur, put some at much greater risk, did so without any warning to those they put at risk, and did so in a way that was biased.

Airbags open with a force powerful enough to harm those who sit within ten inches of the bag or whose fragile body parts—heads, for instance—are at the height of the bag. The first-generation airbags deployed in less than the blink of an eye, at about 180 mph.[58] The bags were designed for "the norm" so that, presumably, it would protect the most number of drivers—those at the norm and, with diminishing effectiveness, those on either side of the norm. People at the norm would be just the right weight and just the right height; so their legs would be just the right length to sit at just the right distance from the airbag so that when it deployed, they would not be so close as to be hit by the airbag as it was inflating, but also not so far away that they would slam into the airbag after it had already inflated. Such "normal" people would have just the right distance away to move forward into the airbag just as it finished deploying so that it would gently cradle them as it stopped their forward motion.

Modifying the explosive force of the airbag, or the size of the airbag so that it could cover a larger area, would not change the fundamental problem: drivers come in such various sizes and shapes that one size cannot fit all of them. The norm was determined by height and weight, and the presumption, apparently, was that they were sufficient to determine the length of a person's legs. The length of one's legs is the variable at issue when we adjust our seat nearer or farther away from the pedals—and thus nearer or farther away from the airbag, on the steering wheel. The assumption that height and weight are sufficient puts to one side tall people with short legs and short people with long legs. The result is that designing for the norm ensures that some are going to be well served and some are going to be put at greater risk. That is an unfairness built into trying to make some safe with an artifact not nuanced enough to protect everyone. Some will be put at greater risk, or at the least not made safer, so that others can be safer.

As we look at individuals who weigh more or less than the norm, are taller or shorter, or who have longer or shorter legs, we eventually reach the ends of the bell curve and find those who are so short or have such short legs that they must sit right up next to the airbag in order to drive and those who are so tall and have such long legs that they must sit far away in order to drive. Those tall, long-legged people will hit the airbag after it has fully deployed and hit it with a fair degree of force, causing harm. The more they weigh, the harder they will hit it. The short, short-legged people will be hit by the airbag as it is deploying, their chest or head getting the full explosive force. They are thus at great risk of being killed by what was designed to save the "normal" person. Their weight will not matter much to how quickly they move toward the bag since they will not have time to move much distance at all, if any, but the larger they are, the more quickly they will be struck by the exploding airbag. So what was chosen as "normal" matters enormously.

The norm chosen was the 50th percentile for men. The airbag was "designed to protect an unbelted adult male at the 50th percentile of body height and weight in a severe frontal crash."[59] The 50th percentile for men was the 95th percentile for women in 1976. So the first-generation airbags protected most men, but a much smaller number of women. More accurately, what protected most men put a good number of women at greater risk—those who are short in comparison to the male norm and those with short legs who must sit close to the steering wheel to drive. The choice of that norm gives new meaning to the principle of courtesy, "Ladies first."

Suppose a 170-pound, five-foot nine-inch-tall male drives his car head-on into a car driven by a 98-pound, five-foot-tall woman. He is as well protected as possible by the airbag deploying in that "severe frontal crash" because the 50 percent male in height and weight was 171.3 pounds and 68.7 inches in 1997 when engineers made their initial choice.[60] She is likely to be severely harmed because of the airbag which would protect her as well as it protects him were she his size and height. It is only because she is so far down the bell curve from the 50th percentile for men of normal weight and height that what protects him can injure or kill her. That at least seems unfair because the engineers' choice is biased against women to the advantage of men.

Yet choosing that norm might be just the thing to do, to minimize the harm to most drivers, if, for instance, males were the drivers in the vast majority of accidents. We would need to look at the evidence about how many males are involved in accidents. Whatever the evidence, we would be making a value choice in determining which group to protect.

Instead of choosing the 50th percentile of males, why not choose the 50th percentile of those who drive? Or the 50th percentile of those who have been in accidents? Are those in the 50th percentile of height and weight also in the 50th percentile of those in accidents? Perhaps smaller men are involved in more accidents than larger men. What about choosing as the relevant group not those involved in accidents, but those who cause accidents? Or what about protecting those who are involved in accidents they did not cause, the victims of accidents? What is the 50th percentile of their height and weight? Or perhaps we should take as the relevant group those who are severely injured or killed in accidents.

Height and weight are the relevant variables for determining the explosive force of the airbag and its size once deployed, but some other variable—e.g., drinking habits—may be better correlated to accidents, and if so, we would then need to determine the height and weight for the 50th percentile of those who drink and drive. Again, we have many possibilities for what may be best correlated to accidents— the age of the driver, the training a driver has either had or not, the driver's use of drugs, both legal and illegal, and on and on.

The engineers had many choices, and any choice is value-laden—as the various examples suggest. Why put shorter women at greater risk of harm if it turns out that they are more likely to be victims of accidents caused by those who drink, say? They then are hit twice, as it were, once by drunken drivers and then again by an airbag whose design puts them at greater risk of injury or death. What is the justification for protecting the 170-pound, five-foot nine-inch-tall male if that should turn out to be the median for those who cause accidents? Why protect best those responsible for the most accidents—if that is the case? In short, no matter what the evidence we would need to tailor our choice of an airbag, any choice we make about whom to protect and whom to put at greater risk is a value choice. Either we value protecting those who cause accidents over those who are victims, or we value protecting those who do not drink and drive over those who do, or we protect women and children first over larger males, and so on. No matter what our choice, we provide greater safety for some at the expense of others, and the criteria for who falls into each of those two groups will reflect a value judgment we have made—consciously or not. As suggested, the engineers might have chosen the median for all drivers, for instance, male and female alike. They may then perhaps have made more drivers safe, depending upon what the evidence shows, and certainly would have saved themselves from the charge of sexism.

To note the obvious, their choice did not produce an error-provocative design. The drivers at the ends of the bell curve did not need to do anything, let alone make a mistake, to be subject to great harm. Engineers are morally obligated to avoid error-provocative designs, but as these two examples show—the Mazda clutch and the airbag—they are morally obligated to cull harmful features out of their design solutions, if they can, whether those features provoke the users into making mistakes or not.

Missed signals and other harms

With some notable exceptions—e.g., the airbags—we have generally used as examples of engineering design solutions problems that we presume are solved by a single engineer and then instantiated, without a misstep, in an artifact to be used by a single operator. That kind of example allows us to isolate out any flaws in the design solution and determine who is responsible for the flaws.

We have thus generally put to one side, for example, engineering projects—e.g., the space shuttle and the problems with the Challenger— where, among other things, responsibility is diffused. The airbag and the Mazda clutch push strongly against these examples of simple artifacts that have been our usual choice, but they make the point that we have focused on only one way in which engineers can make an unethical design choice. Choosing an error-provocative design is an ethical mistake, but, as the airbag and Mazda clutch examples show, it is only one kind of ethical mistake engineers can make in solving design problems. There are many others, and it will be helpful to examine a few more ways in which engineers can introduce harms into their design solutions.

We will stick with the kind of example we have been using—a simple artifact designable by a single engineer where a single operator is to use that artifact. We will be focusing in this section on the signal the artifact sends to the user. The things that can go wrong there mirror all the ways in which we can fail to communicate with someone. We shall examine only a few kinds of failure.

1. **No information**—As we have seen, some artifacts provide no information to the operator when, indeed, they most need it. We have mentioned the automatic faucets that turn on when you put your hands beneath the faucet opening. These malfunction, but nothing tells the people wanting to wash their hands that the faucet is not working. This is particularly galling to those unused to such faucets. They are unable to figure out whether they are doing something wrong and cannot determine what it could be as they look for guidance to others nearby who are successfully washing their hands. A visit to a bathroom in an international airport where travelers unused to such faucets look around for guidance when they are unable to wash their hands provides a memorable experience of just how frustrating it can be for users who find themselves unable to operate an artifact, immediately assume they must be doing something wrong since others are succeeding where they are failing, but cannot for the life of them see what they are doing wrong.

The same kind of problem of an artifact giving us no information when we need it arises for the double doors we find in, say, banks. Typically one door is unlocked—the one on the right as you come in. The other is fastened by bolts at the top and bottom on the inside edge of the door. We go into the bank without any difficulty, but when we try to come out, we get stuck at the door we would normally use—the one now on our right. We get knocked back because it is locked. We learn from experience to be cautious in exiting such places so as not to injure ourselves. The lesson can be quickly learned if you push against the door using your hand with your weight behind it to ensure that the door opens. Your wrist will give way and possibly get sprained. So we learn to push against the door cautiously with our shoulder so that if it is locked, we will not hurt ourselves.

2. **Useless information**—Some artifacts provide useless information. Here is an example received while trying to send e-mail (Figure 7.1):

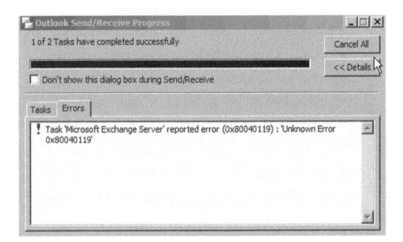

FIGURE 7.1 *Microsoft Exchange Server error notice.*

What is fascinating about this "Unknown Error" is that it has a number: 0x80040119. Presumably other unknown errors have other numbers—which suggests that they can be tracked down and identified, though not by you, the user.

This sort of thing is annoying, to say the least, and software is notorious for problems like this. We are all told we should back up our files. So a friend of mine bought a backup program and proceeded to set it up to back up a bunch of files—a batch backup. When he clicked the "Batch Backup" button, a box popped up which read, "Nothing to back up" with a single "OK" button to click. Not being able to figure out why he was being told there was nothing to back up when he had lots to back up, he clicked the "OK" button, only to have yet another box appear: "Nothing to back up" it read, with an "OK" button to click. He eventually clicked it, and another box appeared with the same message and the same button. He found himself at the edge of an infinite regress.

It turned out, he later discovered, that when the software said, "Nothing to back up," it was referring not to what needed to be backed up, but to what had supposedly just been backed up. So whether you had backed up anything or not, when you click "Batch Backup," the message would appear to be telling you, "Hey, nothing more there to back up!" It is not just unclear, but incomprehensible why the program would stop and tell you that you had completed your task when you had not done anything at all except click the link. And it is incomprehensible why it would then give you a chance to click "OK." Is the "OK" to tell the software that you understand that you successfully clicked the "Batch Backup" button? But then why should it tell you there is nothing to back up?

When someone says something we do not understand, we usually can ask for a clarification. When we read something we do not understand, we can sometimes figure out from the context what must have been meant. But with this "information" from the backup program, we have no idea what is meant and no way to figure out what could have been meant. Worse, we cannot proceed to do what we started out to do, back up our files, because we cannot get past the incomprehensible message that we do not have anything to back up—when we obviously do. Every time we click on the button to back up our files, we are stopped dead in our tracks with incomprehension. The only thing to do seems to be to click on the button again, and that takes us nowhere. So the information provided was useless for my friend—and harmful besides since it was not until long after that he was able to figure out how to get past that dead end in the software so that he could back up his files.

Again, this software example is just that, an example of how an artifact can provide us with information that we cannot use. The way this artifact fails is not the only way an artifact can fail to provide helpful information. Many of us had had the problem of trying to figure out the instructions to put together a child's toy on Christmas Eve, unable to make heads or tails of the "information" provided—with a tab marked "A" to be put in a slot marked "A" when the only slot in the diagram provided is marked "a" and does not seem quite the right spot anyway. And many of us have had the problem, which gets worse the worse your eyes become, of trying to decipher the exceedingly small print that sometimes gets printed on instructions—an offer for an extended warranty with five-point font underneath saying that when you sign, you are thereby authorizing the company to charge your credit card that amount every year. The print provides the relevant information, but the font is so minute that you are likely not even to notice that it says something, let alone be able to read it without searching out a magnifying glass.

3. **Ambiguous information**—We are all familiar with situations where someone says something ambiguous. We take it one way and proceed accordingly when we should have taken it another way. After his loss in the 1962 California gubernatorial election, former Vice President Nixon was asked whether he would run for President. He famously said that he would not run for any other office. Newspapers headlined that he was quitting politics, but what he said was ambiguous. He could just have easily been saying that he had made a mistake in running for some office other than the Presidency, a mistake he would not repeat. And that is in fact what he apparently meant.

Artifacts can give equally ambiguous signals. In one national brand store in the city near me, two sets of doors open to let in customers. Both are opened automatically and so signal to customers that they need not concern themselves with the doors, but can walk in and continue talking, looking after children, thinking about what you are there to purchase. Unfortunately, the second set of doors opens only after you have walked through the first set, and on the other side of those doors, set in the middle about three feet back, is an electronic post that emits a signal if someone tries to leave without paying for something. The post is just high enough to do serious damage if you walk into it, but not high enough to be particularly noticeable to someone otherwise occupied.

Here it is the configuration of these objects that creates the ambiguity. I can attest from my own experience that it hurts to walk into the post. It would have hurt even worse if I had been rushing into the store. I took seriously the signal the automatic doors gave me: "Welcome, come right in! Don't worry about a thing! We'll even open the doors for you!" Placing the electronic post just on the other side of the second door, with no warning to those walking, is similar to enticing someone to do something and then causing them harm if they do—a sort of entrapment. I am not sure whether an engineer was involved in figuring out how to configure these artifacts; we know that even a halfway competent engineer would not have permitted such a configuration.

4. **Other bad signals**—Some artifacts are designed to send us signals. A stop light tells us to stop, go, or proceed with caution, using red at the top, green at the bottom, and yellow in the middle as signs. A stop sign tells us to stop, and as the examples of different stove-top configurations tells us, artifacts signal to us whether designed to do so or not. In the worst of cases, as we have seen, an artifact's signal provokes an error even for the best, those we would presume most adept at understanding such things because they are the most intelligent, the most highly trained, and the most motivated. But there are many other ways in which an artifact's design can mislead.

As we have seen, cars and trucks provide a raft of examples of how things can go wrong. The problems range from an engine so placed in its compartment that one cannot change a spark plug without taking the engine out to a supposed safety feature that will not let you start a car unless your seat belt is fastened. Fastening the belt completes a connection and leaves you at the mercy of wires going through the door that flex each time the door opens and closes and eventually break, leaving you unable to start your car. I had a VW rabbit with that "safety" feature. I found myself one day far from home and any service station, unable to start my car—until I got so upset I slammed the door shut. I sat there and then tried to start the car one last time. It started, much to my amazement. I had apparently reconnected the broken wires when I slammed the door. Scratch a mechanic and you will find a raft of various examples of some engineering mistake or, at least, bad judgment.

   a. **Shoulder harness**—My 1992 Subaru SVX had a harness that is attached to the door so that opening and closing the door opened and closed the harness. Close the door and the harness is across one's chest. The design is awkward since when the door is open, the harness is directly in front of your chest, making it difficult to lift anything out of the car. The design is also dangerous. Both the harness strap itself and the instruction booklet for the car make this point. A tag on the harness strap reminds you to fasten the seat belt or face serious injury, and the instruction booklet says that you may experience "severe head trauma" if you do not fasten your seat belt. Why?

If you do not fasten your seat belt, the harness will impede your moving straight forward in an accident, but nothing prevents you from slipping forward along the seat. As you do slip, your trunk and head will go down to keep up and your chin will catch on the harness. If the impact is severe enough, the edge of the harness can sever your neck.

The problem is that the closing of the harness across your chest makes you feel safe. You are strapped in, after all. I almost always had to ask or remind passengers to fasten their seat belts, and they would inevitably reply, the first time, "Why? What's this?" I would say, "That's the harness. You still need to fasten your seat belt." So you think you are being made safer with the harness when you are being put at greater risk.

One oddity about this arrangement of harness and seat belt is that commonly a seat belt and harness are continuous so that by fastening your seat belt, you fasten your harness. The arrangement in the Subaru does not save any effort. You still need to reach down and fasten your seat belt. That may be another reason why people are puzzled that they still have to do something. It seems unreasonable.

I am not suggesting that the Subaru engineers intended to cause passengers problems with their design. I suspect that they did not think through what complications would ensue from having a combination of an automatic harness and a seat belt that needed to be fastened by hand. The arrangement was unnecessary.

b. **Shower faucets**—Plumbing is also a rich source of misleading signals. Scratch someone who showers, and you will find stories of mishaps. Almost everyone has stories to tell. I am used to shower fixtures that turn on when a knob is pulled out of the main faucet, diverting the water from the faucet to the shower. I found myself in a shower stall in a motel where there was no knob to pull. Indeed, I could not see how to turn the shower on. I then noticed a handwritten sign on the wall next to the shower that said, "Pull the ring down!" Others had also had this problem.

So I looked at the fixtures, hunting for a ring—without success. I then thought that perhaps, like the sign, the ring was outside the tub. So I looked around outside. No ring.

Eventually I found what might be "the ring" at the tip of the tub faucet, but it would not move. Only by sitting in the tub was I able to get a purchase on it, but it would not come down. After much struggling, I was able to turn it counterclockwise. I was then able to pull it down—whereupon I, and the bathroom floor, got drenched as the shower water came on.

This is nice example of at least three different problems we can have with engineering artifacts. As with the Cadillac trunk, nothing about the artifact gives us a clue about how it operates. So the first problem is that we cannot tell by looking how the thing works. Second, we bring different histories to artifacts and thus habits of use that can interfere with our using an artifact that works in a different way than our history has primed us for. That is why, when asked, different people give different examples of problematic plumbing fixtures and why some may strike some of us as not problematic at all.

A third problem is that we bring different bodies to the artifacts. My mother would not have been able to take a shower in that motel. She would not have been able to turn the ring. Perhaps a more telling example concerns the "child-proof" pill containers she found elder-proof. I have no difficulty with them. She had severe arthritis; I do not. The trade-off in designing child-proof containers is that anyone who has difficulties pushing and turning at the same time will find the containers proof against opening them.

Over time, problems that arise in original solutions are generally resolved by modifications in the artifact. The artifacts in our everyday lives are generally themselves the result of evolutionary pressures where problematic designs are weeded out or changed to become less problematic. Pressures work against this evolutionary process, of course.

Companies want to differentiate themselves by having a different product, and progress would halt if we did not constantly push the envelope of design, creating new ways of doing things and so creating new problems to weed out. When the technology moves slowly, we should expect engineers to consider habits of use, for instance, as they push the envelope of design.

The artifact: Sustainability, recycling, and remanufacturing

We have been examining how the artifact signals us and so can mislead us as we try to use it, but engineers also make decisions about the artifacts that realize their design solutions—what it is to be made of, what process will most easily produce the artifact, whether its components can be recycled or remanufactured once the artifact has run its natural life, whether the components can be safely disposed of, and so on. The choices they make regarding the artifact itself are no less moral than those they make regarding its design. Engineers have moral obligations that go far beyond the consideration of those who use the artifacts that realize their design solutions. Those obligations concern the history of the artifact—where the material comes from that is used to manufacture it and how it is obtained to what happens to the artifact after it finishes being useful to an operator. The scope of concern is wide indeed.

Discussion of the moral problems that have produced sustainable solutions requires a book in itself, but we can get a sense of the harms that can occur, and of the difficulties in precluding further harms, by considering the following examples.

1. **Mercury**—Designing artifacts with mercury in them—electrical switches, for instance, or older kinds of batteries, or the current fluorescent light bulbs—puts us all at risk when these artifacts run their natural life spans and are trashed. Mercury is a poison that can harm our brains, our kidneys, and our lungs. The more mercury we use in our products, the more likely it is that we will end up with mercury in our food—in fish, for instance, but also in crops grown on soil contaminated even with small amounts of mercury. Mercury accumulation is cumulative, and so someone with continued exposure from infancy to old age has a greater risk of significant harm. We thus put our health at greater risk than need be were mercury not used. We may end up with more efficient lighting, with fluorescent light bulbs, but at a potentially high cost to us and those who follow us. It would clearly be best for engineers to avoid as best they can designing artifacts which require mercury for their operation.

Mercury in products would not be a special problem were it easy to isolate and easy to recover so that little, if any, escaped into the environment to poison our water and soil. But once mercury is in artifacts, it is difficult to prevent its escape into the environment. We might think of making some entity responsible for its removal from artifacts before it gets into the environment, but the expense is high, and no company that produces artifacts with mercury is likely to accept easily the burden of removing it.

A rather depressing case in point concerns the mercury in the 36 million trunk convenience lights and antilock brakes in vehicles built prior to 2000. Roughly half of these vehicles are General Motors (GM)products, and GM joined a partnership in 2005 to recover the mercury. Since the program began, 2.5 million switches have been recovered, containing 6,500 pounds of mercury. This is a good program, with a good track record, making the companies that produced the risk of mercury contamination responsible for reducing the risk. And it is relatively inexpensive, costing GM less than a million a year.

But that was the old GM. The new GM claims that though "GM's former entity remains a member of the partnership," the new GM "has never produced vehicles with mercury switches and has no mercury switch responsibility under the terms of the bankruptcy court order."[61] We have of course entered into a wonderland here of doublespeak: to say that "GM's former entity remains a member of the partnership" is disingenuous in the extreme since it no longer exists and cannot therefore contribute money. Saying that it "remains a member of the partnership" is like leaving a space at the table for a dead relative: no one will ever show up to share in the meal. "GM's former entity" cannot contribute to the partnership, and in refusing to take responsibility for the vehicles manufactured by "GM's former entity," GM's current "entity" will increase the risk of mercury poisoning for all of us to save less than a million a year.

Still, it is difficult not to admire GM's new entity for its disingenuous way of shifting responsibility so it no longer has to pick up the costs associated with GM products or, rather, products made by a company called "GM" that no longer exists. By calling the former GM "GM's former entity," it is able to keep the name GM and perhaps make use of what good its "former entity" has done in the past while shirking any costs associated with products it—or, pardon us, its former entity—produced. It is easy to figure out why GM's new entity might want to refer to its former self—or, rather, to paraphrase its locutions—the "former entity also known as GM" in that way.

2. **Throwaways**—Designing artifacts which, once their lives end, cannot be used in any way, or only with huge expense, is to condemn us to having trash we must bury with potential harm to our drinking water, air quality, and soil and to the harms attendant on getting the original raw materials and the harms produced in getting more raw materials to make more of these throwaway artifacts.[62] We have come to live in a world full of artifacts that cannot be repaired—toasters and coffee makers that cannot be disassembled to be repaired and for which spare parts are impossible to obtain even if you could take the artifacts apart, wrist watches and cell phones that cease working or are replaced by newer models, cars that begin to go bad after 80 thousand miles or so, and on and on. The list is long.

Indeed, as companies sell more of what their customers need and cannot repair, planned obsolescence may be tempting, among other business strategies, for instance. That ensures that the throwaways must be thrown away that much the sooner, to be replaced with new throwaways, all to the profit of the manufacturers. Few strategies could be worse for us or our environment. This is particularly so when the obsolete artifacts contain harmful substances—mercury, or lead, for instance—or when the cost of removing the artifacts is high enough—old tires, for instance—to tempt more than a few into dumping them wherever they can. We end up with health hazards we could have avoided.

One solution is for engineers to solve design problems in ways that permit the artifacts that realize those designs to be repaired if broken or, when they cannot be repaired, to be reused or remanufactured. Indeed, even if manufacturers continue to produce throwaway artifacts that cannot be repaired, or repaired easily, we can bypass some of the harms involved by designing them so the materials and/ or parts can be salvaged.

For example, by 2011, at least, Mercedes had "a recyclability rate of 85 percent and a recovery rate of 95 percent."[63] The company made a commitment in the early 1990s "to implement a total vehicle recycling program with two main elements: vehicle design and vehicle recycling." The "design efforts...include choosing environmentally compatible and recyclable materials for components, reducing the volume and variety of plastics used, making plastics parts with logos and avoiding composite materials as much as possible."[64]

We might wonder what constitutes the 5 percent that cannot be recovered or the 15 percent that cannot be recycled. It is not the batteries—which is good news for electric cars. It "[t]urns out that the 12-Volt battery is the most recycled product in the world, according to the U.S. Environmental Protection Agency. In the U.S. alone, about 100 million auto batteries a year are replaced, and 99 percent of them ... are turned in for recycling. Roughly 97 percent of the lead in a 12-Volt battery can be recycled. The electrolyte, especially sulfuric acid, can be neutralized, repurposed, or converted into sodium sulfate used in fertilizers or dyes. Even the plastic case can be ground up and reused."[65]

The vehicle design and recycling program Mercedes began was driven in part at least by regulations in the European Union requiring manufacturers to take back what they produced after their useful lives were over, but Mercedes also had financial incentives to do what it did: it saves money to design wisely so that an artifact's parts can be recycled and recovered. If GM's "former entity" had been so prescient, it surely would not have polluted our lives with over 90,000 pounds of mercury.

We do not need to dredge to uncover many more examples of a failure to design for sustainability—a failure to foresee what will happen to artifacts once their useful life is ended (as, e.g., the mercury switches illustrate), a failure to think through how the parts of a product might be salvaged and so put to new uses, a failure to consider the costs of producing the materials from which an artifact is to be made, and so on. The solutions to these problems will depend not only on engineers. We need policies in place that encourage sustainable development, and companies need incentives not to pursue short-term goals of profit over long-term goals less inimical to us and to our environment that would produce more profit later.

We will not proceed with those matters here, but the examples given are illustrative of the kinds of issues engineers need to consider if they are fully to fulfill their obligation not to cause unnecessary harms.

Other harms

The harms engineers can cause are as extensive and varied as the various interests we have. We have interests in living a long life, in living it without serious injury, in being fairly treated, in being able to use technological conveniences without inconvenience, in driving our cars without unnecessary risk, in climbing ladders that remain stable as we mount them, in shaving without cutting ourselves or having our razors burst into flames, and on and on. A harm is a setback to one of our interests, and if the setback is serious enough, the harm raises to the level of ethical concern.[66]

As I have indicated, we will not try here to draw a line between harms that are trivial, even if annoying, and those that raise to the level of moral concern, and we will also not lay out in any systematic way the various kinds of harm that engineers can cause through faulty design solutions or map out the sequence of potential harms an engineer must consider in solving any design problem.

The design solution sits in the middle of that sequence. Whatever choice is made has implications for the beginning of that sequence as well as its end. We can cause harm through how we extract the ore we need for manufacturing and the coal and gas we need to create the energy to extract the ore and coal and gas and power the process of manufacturing the artifact once designed. We can cause harm through how we create substances like various kinds of plastics that we use in our engineered artifacts. We can cause harm in the process of going from a design solution to a manufactured artifact, and we can cause harm in how we handle an artifact once its useful life is over—or, to put the point a better way, how we have designed an artifact so that we can handle it with the least harm after its useful life is over.

The design solution determines the nature of those harms as well as their extent, and since the sequence of events determined by that solution is so extensive, we may think of engineers as stewards of the world. It is their design solutions that determine the features of that sequence and so determine what shall be mined for the resulting artifact, what chemicals will be necessary, the nature of the waste that results—all matters of environmental concern and all matters of moral concern.

As we have shown, some design solutions will cause more harm than others equally effective, and since unnecessary harms ought to be avoided, the choices engineers make about how to solve design problems matter enormously to whether our environment is unnecessarily harmed. That is why engineers are stewards of the world.

Every item in the sequence that leads up to an engineering design solution and every item in the sequence that follows after the solution's realization in an artifact is as much an object of moral concern as the process of going from a design problem to a design solution. Moral considerations enter that sequence at every point. The paradigmatic example of how harms enter the design solution is an error-provocative design, a solution which will ensure that even the best and brightest will cause harm, and we can easily see how unnecessary harms can enter the sequence at the beginning all the way through the end. Any particular sequence is itself an artifact, that is, the result of choices that need not have been made had another design solution been chosen to create a different sequence.

Engineers are to avoid not just those designs that provoke errors, but those designs that are unnecessary and harmful, whether the harms they produce are the result of an operator being provoked into a mistake or not. What matters is not that the operator be an agent in the production of harm—as was the pilot in that Colombia airliner— but that the harms produced by an engineering artifact could have been avoided by a different design solution.

We have considered a variety of harms caused by engineering mistakes, and it is easy enough to find other examples of problems with the big artifacts of engineering—the Hyatt-Regency in Kansas City,[67] the Big Dig in Boston,[68] the flood walls in New Orleans, the Interstate Highway bridge in Minneapolis.[69] As we all well know, things can go wrong with engineering artifacts, sometimes causing great harm. In some cases, the engineers are responsible; in some cases, they are not. But as we have seen, we do not need to look to complex engineering projects to find artifacts engineers have designed that cause harm. We can look, as we have, to the simple artifacts that could readily be the work of a single engineer.

In sum, an engineer is making a wide variety of choices in solving a design problem and must be cognizant of what other choices follow from choosing a solution. The engineer must thus think through various ways in which possible solutions might work when instantiated in an artifact, make the necessary calculations for each possibility, probe the ways in which this or that solution may fail, and trace out the consequences of potential failures to determine which design has the most extensive and expensive failures, which has the worst failure rate, which has the least damaging, which is the easiest and least expensive to manufacture, to ship, and to store, and so on. One job engineers therefore have is to trace out possible design choices to see where they would lead. That requires the engineer to look downstream, as it were, to see what happens once the design solution is realized in an artifact. But the engineer also needs to look upstream, for any design choice has implications there as well. Any design solution entails another set of choices—what materials to use, how much can be recycled or must be lost after the useful life of the artifact, how easy it is to salvage what can be reused, and on and on. Each of these choices can introduce the potential for harm. None is necessarily easy. All are moral choices—as we shall see in looking at a few examples of design solutions.

What counts as a design solution?

Engineers are no different than other professionals in second guessing the solutions they have found for the problems they face as professionals. We often think back over what we did to solve a problem and find something we could have done differently that would have been better. It is at the heart of what it is to be a professional that we presume that we could do better than we did. Writers think about how they could have said something in a different and better way; surgeons think about how they could have cut an operation's time and benefited the patient by using a different method; engineers think about how some alternative solution that had not occurred to them might have been better.

So what counts as a solution to a design problem always carries with it, or ought to carry with it, a tentativeness. We strive, or ought to strive, for the best, and that carries with it the admonition, "Perhaps we could do better!"

As we have seen, it is not difficult to provide examples of design solutions that could have been better. I will provide one here that could have been much better and then provide an example of a solution which strives to be the best. These examples both have to do with signage, a wonderful source of examples for what can go wrong.

1. **Road stripes**—It is a common problem in handling turning traffic at an intersection to have two lanes for turning, and the accidents that subsequently occur are predictable. A vehicle in one turning lane turns into the other lane and hits the vehicle there. Such accidents are most frequent when a new turning lane is added where there had only been one, when drivers do not realize there are two turning lanes, and when drivers fail to see the dotted lines between the lanes. Fender benders are the most frequent result because the vehicles are turning relatively slowly, but they cause harm, obviously, in bending fenders and in creating traffic jams.

One cure is to make the lines between the lanes solid rather than dotted, the signal not to change lanes. That cure still requires drivers to pay attention and also presumes they will do what the law requires, stay in their lanes, but on the presumption that it would help; the highway department for Rochester, New York, did just that for an exit off the interstate. When you take the exit to your right, it is one lane, but that one lane immediately widens into four, and the two on the left go underneath the overpass that carries the interstate.

The highway department did paint a solid line not just between those two lanes but also on either side of those two lanes, preventing—or, more accurately, making it a traffic violation—to move out of the lane you were in, and the solid lines went all the way under the overpass to the road beyond.

I happened to use that exit just as the newly painted lanes were being opened up for use. I was in the lane farthest to the left, and I discovered that far from solving the problem of drivers shifting lanes, the lines had aggravated it.

As is typical for exits off interstates, there is an entrance on the other side, and to enter the interstate from underneath the overpass, a driver has to get into the center lane, the one between the two lanes taking the traffic that has just exited from the interstate and the two lanes going in the opposite direction. But the solid lines to keep vehicles from wandering into other vehicles had been painted so that anyone using the lane I was in ended up in the center lane, the lane reserved for those going onto the interstate. I found myself unable to get onto the proper lane to continue on my way and ended up going back onto the interstate—in the direction from which I had just come.

I could not get out of that lane without crossing a solid white line—a traffic violation, and, in addition, the next lane over was filled with those vehicles that had been in the other left-turn lane. Even if they had realized there was a problem, the drivers in that lane could not move over to the empty lane to their right because they too would have had to cross a solid white line.

I called up the county engineer after I went to another exit and got home. I explained what had happened. He said that they would just wait until the new stripes wore out and then repaint. I suggested that he would not have the luxury of waiting, that the county would be sued long before the paint wore out, and that he might want to go out and look for himself at the problems the lines were creating. He called back later and said that the lines would be painted over in black and new lines would be in place quickly. The problem was solved by the time I next used the exit several days later.

I presume that the problematic solution was a result not of the country engineer having made a mistake in locating the solid lines, but of the painters misunderstanding what they were supposed to do—one of those problems that can occur as a design solution moves to realization. But, in any event, we can put that unfortunate initial solution at one end of a spectrum of solutions where it is easy to see how things could have been done better. Our next example is of a solution that is paradigmatic of how to do things right, using a series of experiments over a long period of time to hone a solution that creates a much safer environment for drivers.

2. **Clearview**—Signage is a continuing source of examples of how things can go wrong. From the one-way signs that face each other to the incomprehensibly complex signs giving information about when a driver can park and when not, signs are a continuing source of misunderstanding. Some of the problems are just what we would expect would infect attempts to communicate—ambiguity, unclarity, grammatical infelicities that create confusion, poor word choice that clouds

the intended meaning, and on and on. In addition, of course, features peculiar to signs can create problems. Those who read them must be able to make them out. Their distance from the viewer, the size of the font, how close together the letters are, the type of font—all these affect the capacity of a viewer to make out what the sign says. If you put an unusual font on a sign, for instance, you are asking for someone to have trouble reading it. We get so used to certain fonts—Helvetica, Times New Roman—that a different font, especially an unusual one, requires more care for us to be sure we have read the text properly, and requiring more care carries with it the increased risk that some will fail to read as carefully as necessary.

Road signs provide some particularly egregious examples of how things can go wrong, and the problems of making out what they say are complicated by the speeds at which those who need to see them are whizzing by as well as by the various capabilities of those who need to see them. Can they read? Can they read fast enough to understand what a sign says? Can they see the sign? We know that one problem is "the amount of light reaching the retina of a healthy sixty-year-old is one-third that of a twenty-year-old."[70] So the dim light that does not bother a young driver may make it impossible for an older driver to see what a sign says. In addition, of course, someone with poor eyesight, even with glasses, can have troubles with glare and light reflected on the glasses that interfere with making out what a sign says. The challenge, then, is to create signage that is readable at a great distance, in different kinds of weather, to drivers with a wide range of capabilities going at high speed.

The Clearview project was a ten-year-long "research program to increase the legibility and improve ease of recognition of road sign legends while reducing the effects of halation (or overglow) for older drivers and drivers with reduced contrast sensitivity when letters are displayed on high brightness retroreflective materials." It also investigated "the ease of recognition of mixed case displays in lieu of all capital letter displays."[71] The project led to the Clearview font and to the use of mixed cases, e.g., "Cincinnati" rather than "CINCINNATI."

The font is significantly different from any of the six different typefaces in Highway Gothic, the official series of the Federal Highway Administration. What is known as the E-modified font has generally been the font of choice from Highway Gothic. "In general," it is said, "the ClearviewHwy lowercase is taller, interior shapes of letters are more open to allow clear definition of each letter, and letter spacing has been designed to accommodate the needs of older drivers when used with both regular and high brightness sheeting materials."[72] Figure 7.2 shows the font's evolution and configuration compared to what has been the standard:[73]

FIGURE 7.2 *Clearview font.*

ClearviewHwy was created by Don Meeker, an environmental graphic designer who got interested in the issue around 1990, and James Montalbano, a type designer who worked on Meeker's original solution. As Montalbano put it, "The fundamental flaw of Highway

Gothic is that the counter shapes are too tiny, ... referring to the empty interior spaces of a typeface, like the inside of an 'o.' When viewed from a distance, and especially at night under the glare of high-beam headlights, the tightly wound lowercase 'a' of Highway Gothic becomes a singular dense, glowing orb; the 'e,' a confusing blur of shapes and curved lines. Meeker puts it more bluntly: 'They look like bullets that you couldn't put a pin through.'" So Montalbano opened the type up, creating more space within the letters. "He understood that Clearview's success would come not from where its shapes are on the sign but precisely in where they are not—the open spaces in Clearview's letters are what make it so readable." [74]

Highway Gothic had never been tested to see how easy, or hard, it was for drivers to make out the font. The versions of ClearviewHwy were tested over and over again to ensure that each iteration was easier to see. As Montalano said, "Signs that you'd be hard pressed to read at 700 feet [in Highway Gothic] were legible at 900 or 1,000 feet," and for a stationary viewer there was "an approximately 40 percent gain, or 200 feet of added reading distance using a 10-inch-high letter on the demonstration panel."[75] A Pennsylvania Transportation Institute study showed significantly increased legibility for an early version of ClearviewHwy. "For drivers traveling at 45 mph, that legibility enhancement could easily translate into 80 extra feet of reading distance, or a substantial 1.2 seconds of additional reading time. On a road with a posted speed of 45 mph, a driver [going at the speed limit] is traveling at 66 feet per second. With Clearview-Bold, the desired destination legend is recognized 1.3 seconds earlier (84 feet) and with greater accuracy, giving the person significantly more time to react to the information displayed."[76]

One crucial insight in developing the typeface is that we more readily recognize patterns created by a mixture of upper- and lower-case than signs in upper-case only. Even if we cannot quite make out the letters, we can recognize a pattern—"Chicago," say, in place of "CHICAGO." The other major insight was that by increasing the height of the lower-case letters, the amount of counter shape—that hole in the "a" for instance—is increased, thus increasing a sign's legibility.

What is admirable about Meeker and Montalbano is that they kept "returning to the font for minor changes: an adjustment in thickness here, a change in letter spacing there. 'Those guys are tinkerers,'" it was said. "They were always playing around, wondering how we could optimize it. We had something we called Clearview, but was there a Clearer-view? Or a Clear-est view?"[77] They assumed, that is, that the font could always be made better and kept working at making it better until they achieved a real breakthrough in legibility.

Highway Gothic, we now know, is surely a less than optimal solution to the problem of making signs legible to a variety of drivers going at high speed, and ClearviewHwy is certainly better because it is more legible and so gives drivers more time to make what may be quick decisions. But we should not presume that ClearviewHwy is the best we can do for road signage any more than we should presume that in painting solid lines for the road stripes—a feature of the design solution that exacerbated the problem the placement of the lines caused—the local highway department picked the best way to encourage drivers not to cross into other lanes when there were two lanes turning in tandem. What the ClearviewHwy project shows, clearly, is that things can be better than they were, and it is that general truth that engineers ought always to presume. A design solution is a contingent choice made at a particular time by a particular engineer or set of engineers, and at another time, with another way of looking at the problem, or another way of testing the results, or with a different engineer or set of engineers working on the problem, or with technological advances, a different and better solution may present itself.

Unfortunately, the US Department of Transportation has decided not to support Clearview, much to the chagrin of some, especially those over 60, who have found the font significantly clearer than the Department of Transportation's Highway Graphic, which is not the preferred font.[78] We can presume that cost was a factor: Highway Graphic is freely available, and Clearview carries a cost.[79]

Value-laden choices

It should not surprise us that our choices can be value-laden. They both reflect and embody values. We express our own values in the choices we make, and some of those values are moral—because of who we are, how we have been trained, what we think or any other personal feature, and so on. Values enter even in seeing that something is a design problem, but I have put that to one side to concentrate on how engineering solutions to design problems— engineering choices—embody values because they are public and can have harmful consequences. In the choices engineers make in solving design problems, we find values embodied if only because the design has effects when instantiated in an artifact that is introduced into the world, and those effects may be beneficial or harmful, or, obviously, both.

Seeing that ethical choices are embodied in the design solutions of engineers ought to change the way engineers look at what they do and so open up new or more careful considerations in design solutions—on what counts as a design problem (so that, for instance, the way a product is to be used is taken fully into account, as it was not for Guidant's defibrillator), on what counts as a harm (so that, for instance, engineers ensure that an artifact's parts can be recycled, that the entire process from manufacture to disposal be sustainable, and so on), and on what counts as a solution (so that, for instance, it is always assumed that things can be made better than they are, as the designers for ClearviewHwy assumed for road signage).

Seeing that ethical choices are embodied in the design solutions of engineers also ought to change our understanding of how to determine the bell curve of competence of engineers. We have focused primarily on what engineers ought to do *at a minimum*. The bottom line is that engineers ought to cause no unnecessary harm. That moral principle is the minimum, however. Causing no unnecessary harm is the least a professional can do, and professionals in any discipline ought to be competent enough not to cause unnecessary harm. But, as we have said before, engineers ought always presume that both they and their design solutions could be better.

8

Role morality

We take on many roles in our lives, and living a role means coming to think in a certain way—as a parent, say, or a lawyer, or an engineer. What marks engineers off are the special knowledge and skills they must come to have, what they need to learn to think like an engineer. These are norms of engineering. No one can be an engineer without learning them, and we properly criticize a failure to track the consequences of a change, for instance, as in the Hyatt Regency walkway collapse, a failure to see how a change reverberates through a design, or a careless calculation with harmful results. These are not just engineering mistakes, but moral faults as well.

For these norms constitute the inner morality of an engineer, what an engineer ought to know to be an engineer, and though we can expect a failure or two just because we are human, a significant failure or a pattern of failures to live up to those norms is properly subject to moral criticism.

In addition, because they are certified as engineers in some way, by graduating from an accredited engineering school, for instance, they take on certain social responsibilities others do not. So a failure to abide by building codes is as much an ethical violation as a legal one.

They also take on a special set of moral relations unique to engineers in taking on a client, and so a failure to follow the norms of the discipline in fulfilling those relations provides an additional moral ground for criticism.

Unfortunately, learning to think like an engineer can mislead us in some circumstances. We will see how that is possible when, among other things, we look at decisions regarding the launch of the Challenger.

Whatever we may wish, we are all born into a social position, a position determined by the nature of the society within which we are born and generally by the social positions of our parents within that society—rich or poor, educated or not, professional or working class, and so on. We thus take on various roles in our birth—the child of middle-class parents, say, a sibling, a member of a large family, or a small one, a citizen, and so on. The number of roles we occupy expands as we grow up. We become a friend—or not—to the neighborhood kids, a student, perhaps an athlete, perhaps an employee as we take on odd jobs, and on and on. Eventually, some will become professionals, and it is the role morality of professionals and especially of engineers that is of concern to us in this chapter. What is true of professionals is true of engineers, and so we begin with a sketch of the role morality of professionals.

At the risk of some confusion for engineers, I will use the term "professional engineer" to refer to those who have graduated from an accredited engineering school (or who have through their experience obtained the equivalent education) and thus can be hired as engineers. A person becomes an engineer—a professional of a certain sort—by going through that process successfully. It is another matter entirely whether a graduate of an engineering school also obtains the certification necessary to become a professional engineer in the eyes of a state and of the profession in general.

On becoming a professional

Exactly what conditions need to be satisfied to be a professional is contentious. Must one be paid? Some professionals work for free, and in some sports, amateurs are paid. Picking out and arguing for any one set of conditions is not necessary to make my point about how ethics enters into professions and into engineering in particular. So the list of potential conditions is longer than those I have picked out as essential. I will not argue that those I have picked are the only essential conditions. These are necessary conditions, but not necessarily sufficient:

1. **A professional must have special knowledge.** A lawyer must know enough about the law to be able to provide good legal advice where "good" means at least "likely to be upheld by a court if things should come to a trial." A surgeon must know anatomy, among other things, so as not to mistake the spleen for the kidney, so as not to cut an artery. An engineer must know about stresses and materials, among other things, so that choosing one material or form of construction will solve the design problem, not create a new one.

2. **A professional has special skills.** A lawyer should learn to argue and parry arguments. A surgeon should learn to handle

a set of tools that require intricate hand–eye coordination and
great care. A surgeon who is into a thrust-and-parry mode of
operation will not survive any longer in the profession than
the patients. An engineer should be able to think creatively
about how to solve design problems, envisage how a design
solution will look once instantiated in an artifact, calculate
stresses and whatever else needs to be specified, learn how
a change in specifications at one point reverberates through
a design problem and solution, and on and on. The rules of
skill anyone should master to become a professional can be
complicated, and what we master are not just rules about
what to do to achieve a certain end—"If you are going to take
out an appendix, this is what you need to do"—but a set of
features that go with mastering those rules. The norms of a
profession are not just the rules of skill that define it, but the
manner in which the rules are applied, the modes of thought
necessary to apply them, and the capacity to tie those rules
together into the coherent whole that defines the profession.
A professional learns to act and to think in a certain way and
cannot just master a subset of the norms of the profession.

3. **A professional must be certified in some way.** A state or
    organization may certify someone as licensed to practice
    a profession. The requirements vary from profession to
    profession, and different states and organizations can
    impose different standards for certification. Sometimes
    all that is required is empirical evidence that the person is
    up to the job. A person used to learn how to be a lawyer
    by being apprenticed to a practicing lawyer. The proof

that the person was a lawyer was the capacity to act as a lawyer would act. Today we generally have law boards, examinations administered by the states to determine if a law school graduate has learned enough to become a practicing lawyer. Other professions also have some certification process. Graduate students must attend seminars to learn the knowledge and skills necessary to their professions and then write a dissertation in their field of study that shows to a group of examiners—faculty members within the same discipline and sometimes from other disciplines too—that the student is capable of producing work at a high enough quality to become a member of the profession. Physicians must not only attend medical school but also work as interns for a number of years, gaining the practical experience that can only come from seeing and taking care of patients. Students who graduate from an accredited engineering school can be hired as engineers and are generally taken to be professionals—although the term "professional engineer" implies, among engineers, that the graduate has also been certified by a state as an engineer, a process that requires more than the completion of an engineering school curriculum.

4. **A professional takes on the potential for a special set of moral relations in becoming a member of a profession.** Professions can in fact be distinguished one from another by the differing sets of potential moral relations they have. A physician who takes on a patient ought to examine the person with great care to see if there are any bodily faults or problems—a probing of limbs and cavities and body parts

for unusual lumps, for instance. The physician has a moral obligation to the patient, and the patient has a moral right to careful care. An attorney who examined a client in such a way would provide sufficient grounds for disbarment from further practice. "But I'm a professional!" would not suffice to get one off the moral hook. You have to be the right kind of professional.

Anyone with a driver's license is familiar with these features. To get a license you need to pass an exam. When you pass it, you are entitled to a license to drive. The exam certifies that you have both the relevant knowledge and the relevant skills to drive and drive safely. You are to know why, how, and when to use turn signals, know the difference between the brakes and the accelerator, know how and when to use the windshield wipers, and on and on. The knowledge and skills we need to drive are no different in kind from those any professional needs, and moral issues enter in the same way for drivers as they do for professionals.

Get in the driver's seat of a vehicle and you take on a special set of moral relations you did not have before. Driving is a risky business. We need only imagine what it is like to be hit head-on by a vehicle weighing several tons and going at high speed to realize just how risky driving can be. That is a risk you take on no matter how good a driver you may be, and any passengers you have take on that risk as well. But because you are driving, they are dependent on your knowledge and skills, and so you have moral obligations to them that you do not have to everyone else. Of course, you also have moral obligations to others—pedestrians, other drivers, and so on. You are operating a heavy piece of machinery capable of killing others and so have taken on a set of moral relations you did not have before.

Not everything that you do wrong when you drive raises a moral red flag, but if the harm you cause is significant enough, you will be morally culpable. It does not matter whether you intended to cause harm or not. What makes you morally culpable is that you fail to use properly the knowledge or skills required for driving safely. If you drive through a stop sign and kill someone, you cannot excuse yourself by saying, "I didn't intend to hurt anyone. I wasn't even paying attention!" You are morally culpable *because* you were not paying attention. You are morally responsible for something you did not intend to do because you would have avoided the harm had you paid enough attention to your driving to stop at the stop sign.

The moral problems that arise for professionals are no different in kind from those that arise for drivers. A professional takes on a special set of moral relations when engaged in professional practice, and moral failures can occur when a professional fails to use properly the knowledge or skills essential to the profession.

As with drivers, only some failures are significant enough to raise a moral red flag. Yet clearly some situations raise moral red flags, requiring investigation—as when a surgeon amputates the wrong leg,[80] removes the wrong kidney,[81] or mistakes a kidney for a gallbladder.[82] We would need to examine these cases in detail to make any moral judgments, but they are troubling just because we have surgeons whom their patients had to trust, but who left their patients far worse off than before—facing life without legs, without kidneys, without a gallbladder.

We do not need to focus on any one profession to understand how ethics enters professional practice. We need only scan the media to find example after example of ethical problems in a wide variety of professions. There is that lawyer in Texas who slept through part of his client's trial,[83] or the coroner in New Jersey who failed to follow the standard procedure of X-raying the victim's skull and so reported death by a blunt instrument instead of death by the two bullets in the victim's head.[84]

Each of these examples involves a professional engaged in professional practice within their own profession, and each causes significant enough harm to raise a moral red flag. We can readily find such examples in any profession. Indeed, the greater the knowledge and the more complex the skills required, the easier it is to fail, it would seem. As we saw, Aristotle said about being ethical, "There are many ways of going wrong,...."[85] In addition, professions are dynamic. Changes are constant, brought on, among other things, by increased knowledge that makes obsolete some of what practitioners may have learned, by technological developments that require new skills, by the continuing refining of old skills and standard procedures, by changes in professional standards mandated by the profession, by legal changes requiring changes in practice.

Such changes can catch practitioners by surprise. I was a Medical Humanities fellow at the University of Tennessee Medical School when the state changed the law to define brain death as death. The medical group I was with had a patient who showed no brain activity upon being tested in accordance with the standard procedure in such cases, a procedure which had been incorporated into the new law. The lead physician told the extended family that the woman was "in a bad way," but that the physicians would do what they could. Once we were out of the waiting room, I asked why the physicians did not declare her dead. It turned out that no one in the group had heard that the state had changed the law. Under the law now in force, the patient was not in a bad way, but dead.

The situations of moral concern are those in which a professional causes avoidable harm. There is a simple, but powerful moral principle at work:

We always ought to avoid causing unnecessary harm.

Given a choice between two courses of action, one of which causes more harm than the other, we cannot justify choosing the one that causes more harm without being morally culpable. If the choice we made is avoidable, that is, we are at fault for causing harm we could have avoided. We determine the harms by examining both the different outcomes and the different ways we are to achieve those different outcomes. If the outcomes are the same between two choices, for instance, but we cause more harm to achieve one than the other, we are morally at fault not to choose the alternative that causes the least total harm.

As we have said, not all the harms we may cause are significant enough to raise a moral red flag, but because the lines will be drawn in different places, for differing kinds of harm, within different professions, it is not worthwhile trying to provide a general rule across professions for what raises a moral red flag. Indeed, figuring out what raises a moral red flag within a profession turns out to be no easy matter. But we shall find ourselves with clear examples of harms that should have been avoided as well as clear examples of harms that do not really matter, and we shall thereby hone in on the crucial lines without laying down absolutely clear markers.

Specifying the details of any role we occupy is not an easy task, and what we shall provide is by no means complete. We shall briefly examine (4) first, put (3) to one side, look at (1) and (2), and then turn to what we may call the form of life of an engineer—looking at the internal morality of that role and then at one kind of problem that can arise for engineers because they have learned to think like an engineer.

Potential moral relations

A careful detailing of each profession would reveal what a person takes on in becoming a member of that profession and in that way how a member of that profession is distinguished from members of every other profession. The knowledge and skills necessary for one profession are distinct from the knowledge and skills necessary for another. A professor needs to learn how to teach, but not how to defend a client before a jury. A lawyer needs to learn how to read legislation carefully, but not how to use a scalpel.

These are features internal to each profession. Additionally, a person who becomes a professional potentially takes on a set of relations, and these potential relations, internal to each profession, differ from profession to profession. In taking on a lawyer, I empower the lawyer to represent me in a court of law if I am charged with a crime. I could hire a nurse, but a nurse lacks the knowledge, skills, and certification to represent me in a court, and no nurse is empowered by the state to act as my representative in a court of law—unless the nurse is also a lawyer. A nurse has an obligation to make sure that a hospital patient is receiving proper care. A professor does not. A professor is empowered to assign students books and articles to read, papers to write, exams to take. A psychiatrist is not. A psychiatrist is obligated to help those with psychiatric problems. An accountant is not. And so on and so on.

These are potential moral relations. A professional is not in such relations just by virtue of being a professional. A lawyer is not empowered to represent me unless I engage that lawyer and become a client or the lawyer is appointed by a court to represent me. So a lawyer has the potential to represent me in court. A physician does not. A physician has the potential to examine me for diseases or bodily harms, but cannot examine me unless I am that physician's patient and so have given consent—or am in an accident, say, and must have a physician examine me even if I cannot give consent. We would certainly look askance, and no doubt sue, if someone came up to us on the sidewalk and, unbidden, started probing us the way a physician is supposed to. "But I'm a physician!" would hardly suffice as an answer to our "Hey! What are you doing?!" The stranger may be a physician, but not our physician, and even our physician would not examine us on the sidewalk.

The moral relations we are examining here that mark out professions one from another are relations a professional takes on in the practice of that profession as the professional takes on patients, clients, and so on. In becoming certified in some way in a profession, a professional takes on the potential to have those special moral relations, and professions are thus distinguished one from another by the potential moral relations those professionals have.

So a person who becomes a lawyer takes on the potential to have the power to represent those charged with crimes. A person who becomes an accountant takes on the potential obligation to use the Generally Accepted Accounting Principles (GAAP) in examining a corporation's or person's books. A person cannot become a judge without taking on the responsibility to hear both parties to a case and to make an impartial judgment. That responsibility is an essential part of what it is to be a judge and in that sense is internal to that position. You cannot be a judge without taking on that responsibility.

A profession is what it is at least in part because of that set of potential moral relations. If I am a judge and you sit down beside me, I remain a judge. You and I have a relation to each other—we are side-by-side—but that relation makes no difference at all to my being a judge. What makes me a judge does not depend upon your sitting beside me, but upon whatever features being a judge requires—being elected or appointed to the office and so taking on the features that belong to that office. If I were a judge, I would carry those features with me whether I sat beside you or not. A judge must have knowledge of the law, a capacity to understand how different individuals could interpret the law differently, a capacity to get far enough into each party's position to understand why each party is willing to go to court to begin with, an ability to assess the merits of legal arguments, and a capacity to back off from the competing legal positions, as it were, and make an objective judgment in accordance with the law.

Another way of putting this is to think of various roles an individual may have—as a parent, for instance, or a sibling, or a citizen in a community, or a physician, and so on. Each of these roles is defined in part by a set of moral relations. Parents have obligations to their children—to feed them, clothe them, rear them well. Citizens have obligations to their communities—to pay their taxes, maintain their dwellings, drive carefully. Physicians have special moral relations that neither parents nor citizens have—to examine carefully those individuals who put themselves in the physician's care to determine whether the individual has an illness that needs curing or some bodily fault that needs repair and then to care for those individuals, doing their utmost to cure them of the disease and to repair their bodily faults. Just so with any professional.

Any one individual occupies a number of different roles—child, parent, sibling, citizen of a community, a state, and a nation, an employee or self-employed (each with its own set of moral relations), a professional of one sort or another or not. Moral tensions occur when one moral imperative in one role someone occupies conflicts with a moral imperative in another role that person occupies. What I am required to do as an employee, for instance, may conflict with what my profession tells me I ought to do. These sorts of conflicts between the different roles we occupy are a rich source of moral problems and are all too often difficult to resolve, the moral imperatives of one role being as powerful and persuasive as those of the other with which it is in conflict. But these sorts of moral problems are not the focus of this book. We are concerned only with how ethics relates to an engineer as an engineer, not an engineer as employee, or an engineer as a member of a team or an engineer as a manager.

We are also putting to one side the moral relations a professional takes on just by becoming a member of a profession. Professionals are at least implicitly licensed by the state, either directly or indirectly because the university in which they received their professional training is accredited. That license exists because the state recognizes a discipline as a profession, and in giving those within that profession a license, it excludes others who are not members of that profession from practicing that profession. That is the reason I cannot legally remove your appendix or represent you in court. What the profession then owes in return for its state-sponsored monopoly on a particular kind of service is that it benefits society in some more substantial way than by simply having its members practice their profession.

In saying this, we are adding yet another set of moral relations to the four we have already articulated:

5. **Social**—By becoming a member of a discipline licensed by the state (because, e.g., the degree is accredited), an engineer has certain social obligations as an engineer and member of the engineering profession.

This set is to be added to the following four:

1. **Role morality**—In becoming an engineer, a person takes on a set of role-specific relations having to do with the practice of engineering, e.g., ensuring that calculations are made correctly.

2. **Design solutions**—The intellectual core of engineering is solving design problems, and because at a minimum, ethically, an engineer ought to cause no unnecessary harm, solving design problems requires ethical considerations if only to avoid a solution which causes unnecessary harm.

3. **External**—When engineers hold an additional position—wearing another hat as an employee, contractor, or manager—they may have ethical problems between what the positions require.

4. **Aspirational**—Because engineering is a profession, practitioners should keep up with the latest engineering techniques, understand how new materials can make for better solutions, be more than merely competent.

Our concern is with the moral relations internal to engineering, but we need to note the other moral relations an engineer may have if only to make it clear that our concern covers only a small part of how ethics and engineering intersect.

Knowledge that

We distinguished above between knowing that something is the case and knowing how to do something. These are different kinds of knowledge. We consider the first in this section, the latter in the next section. But we should note that we do not need to lay out in any detail the knowledge and skills people learn as they become engineers. Engineering books do that. It is worth laying out in more detail, however, some knowledge that engineering students ought to learn that do not usually appear on the list of "things to learn before I graduate."

Engineers need to know a great many things other than, say, how to calculate. One problem they face is that as they work to make things safer, those who use what they have redesigned adjust their behavior to increase their own risk. Feeling safer, the users engage in more risky activity than before. Evidence indicates that ABS brakes have led to no decrease in the number or severity of accidents because, it is suggested, drivers simply go faster. Providing helmets for hockey players has led to an increase in paralyzing neck injuries because, in part, players feel that they can engage in riskier play, if we may call it that, with the helmets than without. Providing helmets for skiers has the same effect. Feeling safer because of their helmets, those 17- to 24-year-old males most prone to accidents go faster and still have harmful accidents. Indeed, paradoxically, equipment that ought to make it safer for some individuals to engage in some activities—driving, skiing—not only fails to decrease the risk they face, because of their off-setting behavior, but also puts others at greater risk than they would have been.

That sort of knowledge of psychology is not an isolated bit, but part of a far broader understanding of how we all engage in activity that engineers ought to take into account to make usable artifacts. The stove-top configurations exemplify how readily we can be misled, and engineers who are not conversant with how we tend to read our environment will choose less than optimal solutions to the design problems they face.

Engineers also need a knowledge of physiology and, in particular, a knowledge of what humans are capable of doing—the norm as well as the extremes. As mentioned, child-proof containers are difficult for many with arthritis and for the elderly who are most likely as a group to be on medication. We approach artifacts with different bodies and differing physical capacities. Solving a design problem in a way that no one is disadvantaged in using it—a matter of fairness and thus of morality—is no doubt an ideal, not easily achieved, but it is an ideal that engineers ought to strive to achieve. As we have seen, we can imagine an engineer purposefully designing artifacts that stymie the best efforts of everyone. "I'd like to see anyone use that can opener without hurting themselves!" We would think such an engineer morally perverse. We would also think morally perverse an engineer who designed an artifact so that it could not readily be used by a particular portion of the population—a door knob that could not be opened except by using one's right hand or a door so heavy and hard to open that only the muscular and fit would open it. Those design choices would be unfair.

Engineers also need a knowledge of the history of the design problem they are trying to solve. In part, this is to ensure that their solution meets the particulars of the problem. It would make little sense to redesign an artifact without taking into account what was causing problems with the previous iteration. Software is a wonderfully rich source of examples for this. For instance, Apple's operating system has built in compatibility with Microsoft's Exchange while Microsoft's Windows does not. Apple saw a problem and fixed it in the latest redesign of their operating system.

But there are two other reasons for engineers facing a design problem to learn the history of the problem and of various solutions. First, there is no sense reinventing the wheel. We can learn from past attempts, sometimes way ahead of their time, as we try to create new solutions. The Selectric typewriter, with a rotating ball rather than individual keys, had its ancestor in one of the first solutions to the problem of connecting the individual strikes on a keypad with making an impression on paper. Blickensderfer designed a typewriter with a removable type ball in 1891. With only 250 parts, versus 2500 for a standard typewriter, it was cheaper to make, weighed far less, was smaller, and had the capacity to type in as many different fonts as there were type balls.[86] An engineer would look very foolish indeed who designed a similar machine and then showed it around, proud of the new creation, only to have someone point out that, yes, it is a good idea, and it was a good idea in 1891 as well. We can look back to those earlier designs and, making use of them as experimental designs, figure out how to improve them. So that is one reason for knowing the history of a design problem and its former solutions—assuming, of course, that it is not a wholly new problem.

Second, perhaps more importantly, engineers need to know what expectations we will carry as we come to the new artifact. The standard example is the QWERTY keyboard, invented to slow down typists so that the keys would not mesh. We could type much faster if the most commonly used keys were placed where they were easiest to strike. We would not then have to use our left-hand pinkie finger for the "a." But changing the keyboard pattern will run against the habits of millions upon millions of typists. Even a single-finger pecker would be nonplussed.

Legacy problems are not morally neutral. We saw this problem when we looked at the Cadillac trunk that closes automatically after being lowered to a certain height: someone, somewhere, is going to do what we are all so used to doing with trunks and so break the mechanism. If we have grown used to something operating in a certain way, and operating a new version in that way will cause harm, then engineers have an obligation to reengineer the new version to ensure that harm will not ensue when users are likely to bring an old habit to bear on a new artifact.

Examples of legacy issues are easy to find, and they illustrate the tension we mentioned when we began this chapter. If engineers are to push the envelope of design, they must be free to change every variable of a previous design solution, but as they change variables, they risk introducing new problems because of residual habits of use even as they try to make things easier for operators. An engineer must thus make informed judgments about how and what to change in solving a design problem.

So, rather obviously, getting the information necessary to make informed judgments is an imperative. Such information is no different in kind than the information engineering students must learn to calculate properly and understand stresses. It is information they need to know if they are to solve design problems in ways that do not cause unnecessary harm. So it is not morally neutral information, but information they ought to have to do their work as engineers.

If an engineer designs a turn-off valve for a water heater made of high impact plastic instead of the old material, lead, we can expect someone who turns it off to give the knob an extra little turn, just as before, to make sure that it is tight—the extra little turn that served to ensure that the lead knob was seated. That extra little pressure may snap off the plastic knob. The valve will then continue on for a bit, opening up so gas can seep out. When the home owner returns home to a basement filled with gas, it will explode when the furnace is turned on. The realization of how deeply ingrained our habits can be, as well as knowledge of how the previous iteration of turn-off knobs worked, is as essential to the engineer choosing the correct design solution as is the knowledge of how to calculate the stresses the knob will undergo when it is tightened in the off-position.

Knowledge how

It is such calculating skills that engineers most obviously need. We have images that come to mind when we think of various professionals—a physician with a stethoscope, a psychiatrist with a patient on a couch, a banker with a cigar, perhaps, and, it used to be, an engineer with a plastic sleeve in a shirt pocket with a pen and slide rule. Now it is an engineer with a calculator of some sort.

That image is not mistaken, but it does not capture the full set of skills of an engineer any more than a physician with a stethoscope captures all the skills a physician needs. There are skills an engineer must have that may not be as obvious as knowing how to calculate. We will mention only a few of the many, but enough to get a sense of how far beyond calculating they extend.

1. **Tracking consequences**—The original plans for the walkways in the Kansas City Hyatt envisioned single rods attached to the ceiling and extending through the "cross beams on which the walkways rested."[87] Those single rods were to be over 45-feet long, and during construction the plans were modified so that the number of rods was doubled and the length shortened, with one set going from the ceiling to the first walkway support beams, attached with nuts and washers, the second set going from the first walkway to the second, again attached with nuts and washers. The walkways collapsed, and 114 people were killed and another 200 injured.

A 45-foot-long rod designed to hold two walkways is not your standard construction item. So it is probably not much cause for wonder that the suggestion to use shorter rods was made or accepted. Henry Petroski quotes a reader of the *Engineering News Record* as saying, "A detail that begs a change cannot be completely without blame when the change is made."[88]

Yet clearly no one thought through the implications of that design change. As Petroski points out, in a telling example, it is one thing to have two climbers on ropes side-by-side going up a stone face and quite another to have one climber on a rope with another climber hanging on to the first climber's legs. The rods that held up the first walkway were also holding up the second walkway.[89] No wonder the walkways collapsed. Indeed, changing the design was not the only mistake made. The walkways would likely have collapsed even without the design change since the original was only 60% as strong as Kansas City building codes required.[90]

In any event, it does not take much skill, after the collapse, to understand the causal implications of the design modification that left the second walkway hanging on the first. But it takes some skill to envisage what changes that modification would make in the way the walkways are supported. It is not necessarily easy looking at the original plans, making the change mentally, and then tracking the projected history of that change and understanding its effects.

That projected history has two aspects. The first concerns how changing one design feature will affect other design features, how the design change impacts other design features and requires changes—how a design change reverberates through the original design solution. The second concerns how the new design solution will play itself out once realized in an artifact. How will the artifact work in practice with that change in place?

This second skill requires a sense of what we may call projective history, of how things play themselves out. That requires understanding the contingencies of life, how completely unpredictable events may occur and change the projected trajectory of an artifact's life. It requires understanding how much space there is between a design solution and the artifact that realizes that design and so how many things can go wrong in realizing the design solution. It requires understanding how people will underuse and misuse the artifact, failing to appreciate, let alone note, its more subtle features and misusing in various ways what they do understand. It requires understanding how even major problems with the artifact may go unremarked, preventing timely corrections.

The collapse of the Hyatt Regency walkways illustrates two of these requirements well. Workman noticed how the walkways vibrated when they used it, but they just worked around it. No one pursued the problem.[91] The engineers who designed the original 45-foot-long rods did not think through how they would be manufactured and how likely it would be for that feature to beg a change. In any event, this skill of projective history is not easy to articulate in all its details.

It is a complex skill having many aspects and many requirements for its proper realization. So it is not easy to figure out how to teach it, how to train ourselves into looking at a design change and seeing how it will play itself out when realized in an artifact. It is no doubt even more difficult when we consider alternative design changes. How will it work out in the long term if this change is made rather than that? Is the artifact more or less likely to break? Is it going to last longer or break down sooner? Will that little change cause problems with shipping and storing it?

2. **Seeing reverberations**—A second skill requires seeing how one design change affects other design features, how a design

change reverberates, as I put it, through the original solution. What else needs to be changed if this is changed? One way of thinking about this skill is to see that it relates to another skill engineers must have in order to complete a design solution.

Decisions have their consequences, as we know, and one feature of any design solution is that any one choice will both open up and constrain other possibilities. Choosing to have booster rockets with segments for the space shuttle, rather than a single tube, created its own problems, for instance. Ensuring that the segments fit together without great risk of allowing hot combustion gasses to escape inevitably leads to something like the O-rings in Figure 8.1:

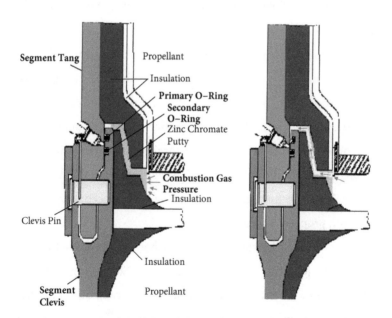

FIGURE 8.1 *Shuttle booster rocket joint.*
Report of the Presidential Commission on the Space Shuttle Challenger Accident (Washington, D.C.: United States Government Printing Office, 1986), p. 57, online at https://www.tech.plym.ac.uk/sme/Interactive_Resources/tutorials/FailureCases/images/CH7Joint.gif

When the rocket is fired, the segments twang, moving against each other and compressing the O-rings—which are supposed to seal the gaps instantaneously. They are made of a substance that will need to withstand the twang at lift-off and rebound to seal any gap before hot gasses can burn through. But they will work properly only if properly seated in the clevis. To ensure that seating, air is forced between them through the valve to the left in Figure 8.1—drawn in the figure, but unnamed, with the nozzle between the primary and secondary O-rings.

Rather obviously, the less the air pressure, the greater the likelihood that the O-rings are not seated properly, but the more the air pressure, the greater the likelihood that the O-rings, and especially the primary O-ring, will be pushed out of their groves. When the engineers at Morton-Thiokol found that hot gasses had blown past the primary O-ring during a launch the January before the Challenger disaster, they tried increasing the pressure to ensure that the O-rings were properly seated, but had no way to determine whether they may have made the problem worse by pushing the primary O-ring out of its seat back toward the rocket booster engine or the secondary O-ring out of its seat out toward the outer casing. No valve was positioned in a way that would ensure that the primary O-ring and the secondary O-ring were both properly seated.

So we have a decision—segmented booster rockets—that has implications for other features of the design. Something like those features seems, in retrospect, inevitable, but those features mean that if there is a later problem, there is no easy way to check to ensure that the O-rings are properly seated.

This skill of seeing how a design decision, or design change, reverberates throughout a design solution is like the skill photographers must come to have of seeing how a seeming minor modification of lighting will utterly alter an image or like the skill that a chess master must have of seeing how a pawn move made now will affect play ten or more moves farther along. We cannot depend upon our intuitions here, but must learn how to see such implications.

Part of that learning will come from practice. It is easy enough for a photographer to take photo after photo, under differing light conditions, until the effects of changes in lighting become obvious, and it is easy enough for chess masters to learn from playing and reading many games and varying moves to see their ripple effects through later stages of the game. It is somewhat more difficult for an engineer to practice building walkways to see how best to support them, but somehow the skill to track the changes of any particular design change must become obvious for the engineer.

3. **Enormous care**—One additional skill we should emphasize is the great care that must be taken in calculating the various components of a design solution. It is a mistake to think that engineering consists in calculating, as though all engineers do is reducible to what they can do with a calculator. Even if that were all they did, they must do it with enormous care, and that skill is no easier to master than any other of the skills engineers must have. It is easy to think a calculator is always right, but the programs engineers use in their computers for calculating are as prone to problems as any other software

program. These programs are artifacts too and are subject to laws just like any other design solution.

It has been argued that

changing a seemingly innocuous aspect of an experimental setup can cause a systems researcher to draw wrong conclusions from an experiment. What appears to be an innocuous aspect in the experimental setup may in fact introduce a significant bias in an evaluation. This phenomenon is called measurement bias in the natural and social sciences.

Our results demonstrate that measurement bias is significant and commonplace in computer system evaluation. By significant we mean that measurement bias can lead to a performance analysis that either over-states an effect or even yields an incorrect conclusion.[92]

Engineers thus need at the least to learn what would count as the right sort of answer to a calculation, one in the ballpark of answers. Before the advent of computerized cash registers that tell the cashier how much change to give, I was handed change more than once that was more than I gave. The cashier was clueless as to what would constitute ballpark change and so had no idea there had been a mistake. But computerized programs introduce their own problems. We use them and assume we get the correct answer without a ballpark sense of what the right answer should be. That can lead to our not catching mistakes.

We know that any one mistake can throw everything off. An error in calculation, for instance, can reverberate through the entire enterprise, throwing off everything else, and it is all too easy to make mistakes—especially in complex engineering projects. The mistake that doomed the Mars Climate Orbiter spacecraft was that one team used "English units (e.g., inches, feet and pounds) while the other used metric units for a key space-craft operation. This information was critical to the maneuvers required to place the spacecraft in the proper Mars orbit."[93] So engineers must not only make the right choices but be sure they get the calculations right so that harm does not sneak in because of an error. The Orbiter error was in part due to lack of communication between the teams working on the project, but that just illustrates one more failure to take due care and check to be sure both teams were using the same units of measurement.

Seeing how a change reverberates through a design, tracking the consequences of a design solution once realized in an artifact, taking great care at each and every step—these are only a very few of the set of skills an engineer must master. They also need the capacity to see problems that call for an engineering solution, the skill to analyze the problem into manageable parts, an imagination sufficient to sketch out alternative solutions to the problem, a grounding in the possible to understand what could work and what would not, what could readily be realized in an artifact, what could not, and on and on.

An engineer is thus a marvel of numerous skills of a wide variety, far beyond the ability to calculate. Merely to become minimally competent requires far more than being able to calculate stresses, or measure rods, or understand how to fit pieces of an artifact together without introducing structural weaknesses. Just learning how to analyze a problem without leaving out any crucial piece of the puzzle is difficult enough, but an engineer must also learn to imagine alternative solutions so as to be more sure that the selected design solution is the best available. Part of the role of an engineer is thus, at a minimum, to have a set of skills, both numerous and varied.

Forms of life

We take on a role in becoming a professional, and that role makes demands upon us that we cannot ignore without putting at risk our professional standing—not necessarily in the sense of losing our accreditation, but certainly in the sense of our being respected within the profession as someone competent. I was once told of a physician who did all the circumcisions at one of a city's hospitals, three or four hundred a year. He tended to take too much skin, causing the boys much pain as they grew and ensuring that they would need another operation for a skin graft. No physician who knew of this was willing to testify as to what the physician was doing, but, to put it mildly, the physician was not a respected member of his profession.

Being a professional demands, at the least, that one be competent, and that means that we come to know all we need to know to be a professional within our chosen discipline, that we come to have the skills we need to have as a professional, and that when we take on clients, we do what the moral relations we have then taken on require us to do.

But if we consider only these demands, considering them one by one and see what we get at the end, we will not succeed in understanding fully what it is to become a professional. Becoming an engineer, for instance, does not mean just learning physics, calculus, and so on, the kinds of things engineers must learn as distinct from those a lawyer must learn—the relevant law for the state in which the lawyer practices, the proper forms that need to be filled out, and so on. Becoming an engineer does not mean just learning a set of skills— as though it were enough to know how to manipulate numbers or run particular computer programs. To become an engineer, to become any professional, is to enter into a form of life that is distinctly different from any other.

The person who acquires the special knowledge and skills of a profession learns to think in a certain way, for instance. Jokes about individuals in various professions depend upon this fact about becoming a professional. The joke about the priest, the physician, and the engineer about to be guillotined displays well the way engineers are trained to think. The priest is led to the guillotine, the cord is pulled, and nothing happens. "A miracle!" exclaim the executioners, and they let him go free. The physician is led to the guillotine, and the guillotine again fails to drop. "Another miracle!", and he goes free. The engineer looks up as his head is being put under the blade and says, "Wait! Wait! I think I see the problem!" That engineer has a form of life, and in particular a distinct way of thinking about the world that will make his life shorter than it would otherwise have been. Yet learning to think that way is a necessity for someone who wants to be an engineer.

Learning to think in a certain way is necessary to become a professional even if it can also be a problem. You need to learn to think like a lawyer, like a philosopher, like a physician if you are to become a lawyer, a philosopher, a physician. The joke illustrates one kind of disadvantage. If we approach all life's problems with only one way of thinking about the world, we may find ourselves creating new problems for ourselves.

These forms of life are not morally neutral, that is. Indeed, a form of life can even be in moral tension. A profession you enter may require competing modes of thought. A colleague of mine at a major Big Ten university heard tittering outside his door one day. It went on, and finally irritated, he went out to tell whoever was doing it to stop. He found that the sound came from a room down the hall where twenty second-year medical students were learning how to examine patients by examining each other. These interns had gone to the same classes together for a year and a half; some had dated; some were dating; some wanted to date. They were learning how to look at a nude body without getting embarrassed and so getting red-faced or, worse, laughing, a typical response to embarrassment. The last thing you would want when seeing your physician is to have the physician, upon seeing you for an examination, double up in laugher and, red-faced, say, "Sorry." It is one important feature of becoming a physician that you learn how to examine other human beings, and that means seeing them as, say, a mechanic sees a bicycle, as a mechanism that may need repair.

Yet we also want physicians to have a good bedside manner, to be able to relate to you not as a mechanic to a bicycle, but as a person to a person. So those training to be physicians need to learn how to switch back and forth between two different ways of looking at their patients. That is not easy, and one consequence is that we have many physicians with poor bedside manners. For some reason it may be easier to see patients as mechanisms than as persons, or perhaps those who successfully navigate the long and hard medical training are those who are best able to see patients as mechanisms.

In engineering, we find a similar set of mixed modes of thought. We demand that engineers learn to think quantitatively and that they be risk-averse. It matters in working out the details of a bridge truss, for instance, to get it right, to do the calculations that ensure that the bridge will not fail because of a truss that is not strong enough. To ensure that the bridge will not fail, engineers are to ensure that the truss is stronger than it needs to be, to be risk-averse, that is. We also want engineers to be creative in solving the design problems they face, to look for new ways of doing things, to push the envelope of design. These two different modes of thought are not incompatible, but they are certainly in tension. Pushing the envelope of design is to risk failure.

The Tacoma Narrows Bridge is a case in point. The design was

unconventional... in that the depth of the roadway structure was diminished by employing a stiffened-girder design rather than the then-customary and necessarily deeper open truss. This innovation gave a slender silhouette whose appearance was dramatic and graceful, but the shallow, narrow span was also very flexible in the wind.

The bridge became known as the "Galloping Gertie" because it "undulated uncontrollably" in the wind. "Subsequent analysis of the Tacoma Narrows failure confirmed that the bridge span acted much like an airplane wing subjected to uncontrolled turbulence."[94] Pushing the envelope of design solutions pushed the risk of failure too high.

Taking on the form of life as an engineer is thus to enter into a life of tension between competing imperatives. Engineering is no different than any other discipline in this respect—as we have seen from the example of physicians. It is also to enter into a particular way of thinking. The joke about the engineer about to be guillotined is funny, and telling, because it is a commonplace for professionals to see everything in terms of their own profession. It may, or may not, be an exaggeration to say that prosecuting attorneys are always prosecuting (though that might help explain their higher than average divorce rate) or that economists see every issue as economic.

We want professionals to become so good at thinking in a certain way that it becomes second nature: they never have to think about how to think about a problem because thinking of a problem as an engineer, say, or as an economist comes naturally. But the better we are trained into a mode of thought that becomes second nature, the less likely we are to realize that we are embedded within a particular mode of thought. An Englishman once said after he circumnavigated the globe that he was happy to be back in England where people spoke the way they thought. Becoming so used to thinking in a certain way that it is second nature risks blinding us to the contingency of that mode of thought. Thinking in English is as contingent as thinking like an engineer, or like a prosecuting attorney, or like a physician, and that is a problem when, like the Englishman, we become so arrogant about our particular form of thinking that we dismiss other forms of thinking.

We can become blind to alternative modes of thought and so fail to see a problem in all its richness or fail to realize that others are reasoning differently, but perfectly appropriately, given the mode of thought into which they have been trained. Modes of thought elevate some relevant features of a full understanding and depress others, and so, as we shall see, modes of thought are not morally neutral. A failure to understand the limitations of the form of reasoning we have been trained into so as to become professionals within a discipline can lead to confusion and moral problems. Some of the decision-making the night before the fatal launch of the space shuttle Challenger illustrates the issue well.

The Challenger and Mr. Lund

The shuttle booster rockets consist of sections placed on top of one another—as shown in Figure 8.1. The problem that mode of construction creates is that the hot combustion gases can escape at the section joints when the propellant fires and the rocket vibrates at lift-off. The solution was to provide a clevis and tang joint with two rings—the primary O-ring and the secondary O-ring. When the rocket vibrates at lift-off, the O-rings will be compressed, and if they were not resilient, they would not bounce back quickly enough to fill the gaps their compression would create. That would allow hot gases to escape and burn through the side of the booster rockets. The rings are made of a rubber-like resilient material, Viton, and are to bounce back into shape quickly enough to preclude the hot gases blowing by.

The evening before the launch of the Challenger shuttle, NASA had a teleconference with the contractor for the shuttle booster rockets, Morton-Thiokol. The overnight temperature at the launch site was predicted to go as low as 18°F, and Viton was certified only down to 25°F. NASA's concern was that the O-rings would become too cold to retain their resiliency. The previous January, when the O-rings were calculated to have been at 53°F, significant blowby occurred, with the first O-ring highly compromised and soot deposited on the second O-ring. NASA's query was whether the O-rings would remain sufficiently resilient after the shuttle booster rockets had been subject to an overnight low of 18°F so that the Challenger could be launched at a temperature of 28°F, the projected temperature at time of launch.

The engineers at Morton-Thiokol all agreed that no launch should occur. The risk of catastrophic failure was simply too high, with the temperature of the O-rings far below the 53°F calculated for the O-rings the previous January. How far below is difficult to tell.

To determine the temperature, we would need a calculation similar to the one that produced the 53°F figure for the launch of the previous January. That calculation would depend upon how cold it got the night before the launch, how long the booster rockets sat outside in the cold, how much heat the booster rockets retained from before the cold snap, and so on. The engineers could not possibly have determined that the night before the launch. It was sufficient for their concerns, however, that the O-ring temperature would be found to be *significantly* below 53°F—whatever the exact figure may have been. The engineers and managers were unanimous in their recommendation.

NASA insisted that Morton-Thiokol revisit that recommendation. We need not get into the complex details of the exchange that followed.[95] The crucial point for us came when Mr. Mason, senior vice-president for Wasatch Operations at Morton-Thiokol, said that it was time for a management decision. He said in an interview that the discussion at Morton-Thiokol had gone on to the point where people began repeating themselves. As Thiokol's Bill Macbeth put it, "when you get that kind of an impasse, that's the time management has to then make a decision. They've heard all of the evidence. There was no new evidence coming in, no new data being brought up, no new thinking, no new twists being put on it from our previous position, and we were just rehashing. And so Mr. Mason then said, 'Well, it's time to make a management decision. We're just spinning our wheels.'"[96]

He asked the engineers if they had anything new to say, and when no one responded, he said that he supported a launch decision and turned to the managers, asking them, one by one, for their opinion. Two recommended launch, but Bob Lund, the vice-president for Engineering, hesitated when Mason turned to him. Mason then said to Lund, "It's time to take off your engineering hat and put on your management hat."[97]

What has become the standard view of this is that Mason was asking Lund to look at the problem from the point of view of a manager, doing what managers are supposed to do, taking a different—and wider—perspective than that open to engineers. When we are well trained into a discipline, the discipline's way of thinking becomes so natural that it may never occur to its professionals that it is only one way of thinking about things, that a form of thinking is a contingent matter that may be inappropriate in some situations for some kinds of problems. So on this standard view, Mason was asking Lund to stop thinking like an engineer and think like a manager. That perspective was different from that of the engineers in at least two ways:

> First, engineers would not normally include in their calculations certain risks – for instance, the risk of losing the shuttle contract if the launch schedule were not kept. Such risks are not part of their professional concern; such risks are properly a manager's concern. Second, engineers are trained to be conservative in their assessment

of permissible risk....Engineers do not, in general, balance risk against benefit. They reduce risk to permissible levels and only then proceed. Managers, on the other hand, generally balance risk against benefit. That is one of the things they are trained to do.[98]

The engineers were looking at the risk to the shuttle from the cold, but the managers were to look at all the risks, including the company's responsibility in delaying a launch and so putting at risk the company's shuttle contract.

If we "balance risk against benefit," taking into account the likelihood and magnitude of each harm occurring and the likelihood and magnitude of each benefit, the risk of a catastrophic failure of the shuttle is just one of many, and the risk is small, the shuttles having launched many a time with a risk of blowby and having flown successfully. Using a risk/benefit analysis, the choice to launch is not hard to make.

Thinking like a manager is very different from thinking like an engineer. As the now standard reading has it, Mason was asking Lund to stop thinking like an engineer and to think like the manager he was. That meant using a completely different decision- procedure to decide what to do. Mason was asking Lund, "Don't think about minimizing potential losses, but calculate the likelihoods and magnitudes of potential losses and gains and then compare the two, letting the results of that calculation determine what to do." Think like a manager!

What the Challenger example illustrates is an all too common problem. We are trained into a discipline and so learn to think in a certain way and carry that with us as we approach other problems. That is the point of the story about the engineer about to be guillotined who stops the proceedings so the problem can be fixed. Short-sighted? Yes. But we smile in recognition because the problem is so widespread. I learned as a philosophy graduate student to take arguments, including my own, pin them to a wall, as it were, and then critique them mercilessly: "Well, the first premise is ambiguous and false on either interpretation, and the argument is not valid in any case." It took awhile for me to learn that my colleagues in committee meetings—especially from other disciplines—did not appreciate the favor I was doing them by pointing out the flaws in their arguments.

This sort of problem is internal to each profession. Lund later said to Chairman Rogers of the Presidential Commission that he had not realized that he had changed his way of thinking when he changed hats. He said that they had "always been in the position of defending our position to make sure we were ready to fly," but that they "changed their position ... to prove to them that we weren't ready." That meant proving to them that the "motor wouldn't work"—presumably meaning that the engines would fire, since that they would not was not at issue, but that the O-rings would not seal. In any event, the managers could not prove that the O-rings would not seal.[99]

Lund's explanation does not quite capture all that happened, however. The rest of his story must be, according to the standard reading, that if they cannot prove that the O-rings will not seal, then they were free to consider their not working as one risk among many. The managers moved from the engineers' position of being risk-averse and so hedging their bet to minimize potential losses. Once they moved from that position, they had only two choices. They could show that the O-rings would not seal, that it was not an issue of risk at all, or they could consider their not working as just one risk among many, do a cost/benefit analysis, that is. They judged, as managers, that the cost of not launching was high enough to trump what they judged to be a low risk of catastrophic failure.

It is not clear what Lund would have done had he realized that he was changing hats and so changing the decision-procedure he was using. Had he realized that, we would hope that he would have asked the obvious question, "Which decision-procedure is the right one to use?" That is not necessarily an easy question to answer, but it needs to be addressed.

Decision-procedures are not themselves morally neutral. If I suggest we flip a coin for something—"Heads you win, tails I win"—I am suggesting that we use a decision- procedure for determining who gets something, and the procedure presupposes that neither of us deserves it. If I take something of yours, and when you realize I have taken it, I suggest that we flip for it, you will quite reasonably be flummoxed at my suggestion. It is yours, after all, and why should you let the toss of a coin determine whose it is when it belongs to you? Flipping a coin to determine ownership only makes sense where neither party owns the object in question (and no one else does either).

Just so, a cost/benefit decision-procedure is as morally loaded as one based on minimizing potential losses. We need not get involved in a detailed analysis here of those differences to see that they are different, with different outcomes that have moral consequences. The cost/benefit analysis of the managers put the astronauts at risk; a decision based on minimizing potential harm would not have done that.

As necessary as it is to learn to think like an engineer to be an engineer, not realizing that other disciplines embody other modes of thinking that give different results when put to a problem can be an enormous disadvantage, as Mr. Lund discovered. Unfortunately, learning to think like an engineer, or like a physician, or like a lawyer is difficult enough. We need to train ourselves out of whatever had been our standard way of looking at a problem. We should want the mode of thought of our discipline to become second-nature to us, such a natural way of looking at problems that it would never occur to us, even as newly minted professionals, to think about them in any other way. Our aim is that we never have to think about how to think about problems in any other way than as an engineer, say. The last thing we should want if we are in a group of engineers looking at a bridge at risk of failure is for the other engineers seriously to consider options that are outside the parameters of an engineering solution. Putting up a sign saying "Bridge may fail!" is not a serious engineering option. Preventing the bridge from failing is the problem the engineers are trying to solve, and it will take engineers thinking of engineering solutions, not sign painters painting a sign, to solve the problem.

Yet if the aim is to train ourselves into a mode of thought that becomes second-nature, it may seem paradoxical for us to try to get out of our own skins, as it were, and come to see that we have been trained into a mode of thought that other disciplines may neither share nor understand. Thinking about how we are thinking requires a special set of skills, but we can only come to recognize the contingency and moral implications of any particular mode of thought by getting distance from what has become a natural mode of thought for us and understanding that it is only one among many—just as thinking in English is only one possibility among many. If we aspire to be the best in our chosen profession, we must learn to understand both the strengths and the limitations of its natural mode of thought. Thinking about a problem in only one way is a recipe for disaster.

That is what Mr. Lund's change of mind tells us. Had Mr. Lund fully understood and appreciated his role as a manager, he would have been able to see how differently he was being asked to think when Mr. Mason asked him to put on his manager's hat, and had he also fully understood and appreciated his role as an engineer, he would have been able to see how the differing modes of thought of these two roles were in conflict. He might also have been able to see that the conflict cannot be resolved just by choosing one over the other. It can only be resolved by considering the reasons for adopting one mode or the other in the situation in question.

Inner morality

Ethical considerations thus enter even into the way engineers think. The mode of thought into which they are trained is not itself morally neutral—any more than flipping a coin is a morally neutral decision-procedure. But as we have seen, the form of life of an engineer is far richer than this examination of their risk-averse decision procedure may make it appear, and ethical considerations appear in many ways, some no doubt decidedly unexpected.

Let us return to the sort of problem we examined in Chapter 2, an accident that requires analysis for us to understand how to prevent a repetition. It is not an accident involving engineers, but out of this accident we shall draw a few lessons about the inner morality of professionals and of engineers in particular.

In a 60 Minutes Report on March 16, 2008, Dennis Quaid and his wife Kimberly reported on the near death of their newly born twins. The twins had been taken to the hospital a few days after coming home because they showed signs of a staph infection.

Part of the standard treatment in such cases, apparently, is the use of a blood thinner to prevent clotting. But the twins were given a blood thinner that turned their blood to the consistency of water. It was pouring out of them, leaking out everywhere it could—their belly buttons, their noses, their toes.

Kimberly Quaid had a premonition that something had gone wrong, and so Dennis Quaid called the hospital at 9 p.m. to ask if the twins were O.K. He was told that they were, but, in fact, they were not. When the Quaids came to the hospital the morning after the disaster, they were met by their pediatrician, the head nurse, and a lawyer from "risk management,... the liability division of the hospital."[100] The Quaids had not been called about the problem.

The twins should have been given Hep-Lock, a blood thinner for infants. They were given the adult version, Heparin. They should have gotten 10 units; they got 10,000—a thousand times more than prescribed, and they got it at least twice. The President of the hospital where this occurred said of the infants' near-death experience:

"It was the result of human error."

The spokesperson for Baxter, the manufacturer of the two blood thinners, said it was not their fault:

"The errors that the hospital has acknowledged were preventable and due to failures in their system."

Both these statements blame the operators, three in this case according to the hospital: the individuals who put Heparin in the drug cabinets for the nurses to use, the person or persons who took the drugs from the cabinet to give to the nurses, the nurse or nurses who administered the drug.

The spokesperson for Baxter said that the way to prevent such errors is to read the label, but the containers for Heparin and Hep-Lock are very similar and easy to confuse. They are the same shape and the same size, with labels differing only because one is a slightly darker blue than the other with a different name, but all in the same font (Figure 8.2):

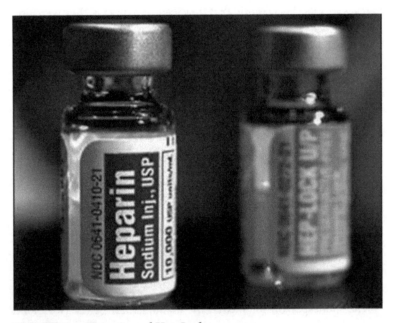

FIGURE 8.2 *Heparin and Hep-Lock.*

The previous year, six infants in Indianapolis were also given Heparin instead of Hep-Lock. Three died, and as a result, Baxter sent out a warning and redesigned the container for Hep-Lock so that it was visually different from the container for Heparin and required the removal of a plastic cover. What Baxter did not do was recall any of the old stock, and the Heparin that caused the near death of the Quaid twins came from old stock.

It would be naive to accept the statements of Baxter and the hospital President at face value. Neither can be read straight, as statements of fact. Both are attempts at risk management, Baxter explicitly saying that it was not their fault and both saying that it was a preventable human error—"not our fault," that is, but "operator error," the fault of those careless nurses and others in the hospital.

It would also be a mistake, however, to dismiss the claims. They purport to be truth-carriers, and so we should do what we need to do to determine whether they are true. To determine that, we would have to investigate in some detail, as the 60 Minutes Report did, exactly how it came to be that the drug was administered to the infants. We would need answers to the three questions we must ask regarding any accident:

What about the operators? Intelligent, well-trained, motivated to do the right thing? Were the nurses and others involved capable?

What about the situation? Unusual, different enough to cause problems even for the most well-trained, intelligent, and motivated operator? Did the problems occur at the end of a particularly hectic day? At the end of a shift?

What about the object at issue? So well designed that it would ensure that no operator error could cause something untoward?

Only after answering all three questions will we be in a position to determine the truth, or falsity, of the claims being made. In the case of the Quaid twins, it seems both that some in the hospital made mistakes and that there is a reason for their making mistakes, the objects at issue—the containers for the drugs—making it more likely than not that even the most intelligent, well-trained and highly motivated individual would mistake one for the other of the two medications at some time or other. The design all but guaranteed that an accident would occur, but only with more detailed information will we be able to provide a properly nuanced judgment.

Yet some moral judgments are easy to make:

- Baxter should have recalled all its former stock so that the kind of accident that occurred in Indianapolis could not occur again because of any failure on Baxter's part to ensure that the drug containers were clearly distinguishable.

- The President of the hospital had no right to accuse anyone of human error without a full investigation.

- Baxter was in no position to accuse the hospital of "failures in their system" both because Baxter itself had failures in its system that created the conditions for the disaster and because it made those judgments without doing a proper analysis of the problem to see if it was in any way responsible.

- Those in the hospital responsible for the twins were wrong not to inform the Quaids when they called that there was a problem with the twins.

- And it was impolitic in the extreme to have a hospital lawyer at the door to the twins' room when the Quaids arrived: it sent the signal that those in the hospital were far more concerned to limit the hospital's liability than to help solve the problem and save the twins.

These moral judgments are judgments about particular acts and omissions by Baxter and by those in the hospital, and none require any significant analysis—though each is tentative, of course, refutable by additional evidence or good reasons for what all the world look to be unethical acts and omissions.

These moral judgments are easy to make in part because of the relations we take up with others once we occupy a particular role. In taking on the Quaid twins as patients, the physicians and nurses at the hospital took on moral responsibilities to the twins and to the Quaids, responsibilities the administrators in the hospital have an obligation to support and encourage. Those in the hospital breached those responsibilities when they failed to inform the Quaids of the problem when they called. These are responsibilities that are functions of the hospital having taken on patients. But the manner in which those in the hospital responded to the problem the improper medication caused puts into doubt the integrity of those who responded and those responsible for those responses, and it puts at risk the integrity of everyone else in the hospital. Any potential patient would have to wonder whether the problems the Quaids had was a fluke, uncharacteristic of care at that hospital, or the uncovering of a systemic problem—a feature of the hospital's character, as it were.

Why did those responsible not inform the Quaids when they called? It is difficult not to believe that they hoped to take care of the problem without the Quaids ever finding out. Why did those responsible have a lawyer by the door waiting to receive the Quaids the next morning? It is difficult not to believe that those responsible had a paramount interest in limiting their liability. This is not the sort of behavior we expect from someone of good character, and it is equally not the sort of behavior we expect from those in a hospital—an organization whose stated purpose is to provide care for the sick and whose employees are supposed to act to further that stated purpose. Through their actions, those responsible are like a physician's saying in such a situation, "I do not tell the truth when it might cause me harm" and "I am also more concerned about being sued than about helping my patients." We would think that, at the least, the physician had a character flaw.

Something more is wrong, that is, than just a single unethical act or omission. Consider Baxter's response. Baxter is a pharmaceutical company. It makes drugs which it then sells so that patients can get proper medication. Patients and medical care professionals cannot know, and have no easy way of finding out, whether those drugs are properly made, whether they are what they claim to be, whether Baxter has taken due care in manufacturing them so that they always have the same amount of ingredients, with their ingredients thoroughly mixed, in containers properly marked with the right ingredients of the right size. In short, we must trust that those in charge at Baxter have done what its selling drugs obligate them to do. They failed to do that when they did not recall the former stock. They put children at risk and did so knowingly, aware that the problem of packaging had caused three deaths already.

It is difficult not to believe that those in charge at Baxter put profit over the potential harm that it knew its packaging could cause, that they did just what those at Guidant did when they discovered that their implant could short-circuit: they traded the company's reputation for money.

Those in charge at Baxter and Guidant and the hospital did not just make a moral mistake. The way in which all responded to the criticism they received indicates a deeper moral problem. Each pointed the finger of blame at others, trying to deflect criticism from themselves rather than trying to determine what went wrong and fixing the problem so that such harm could not happen again. Because of that sort of response, we have another sort of moral problem: those at these companies and the hospital have lost their moral compass. Would you trust a Baxter representative who told you, after yet another "accident," that it was not Baxter's fault? Guidant? The hospital? We all make mistakes, and we can excuse even the most grievous of errors if those making it respond appropriately.

But these three responded in a way that puts their corporate character in question. They responded the way a restaurant I was patronizing responded to a complaint about a fly in a friend's rice— "flied rice," as another friend called it—by blaming the waiter rather than apologizing and saying they would work to ensure it did not happen again.

What is even more morally appalling—and the reason these examples have been chosen—is that those at Baxter, Guidant, and the hospital all have taken on special obligations to help by being in the businesses they are in. Baxter manufactures and sells drugs that are to help the sick; Guidant designs, manufactures, and sells heart implants to those whose lives depend upon electrical circuits firing to restart their faulty hearts; the hospital is licensed to care for its patients. All three betrayed that obligation to help when they responded by blaming others rather than by investigating what went wrong. We want to say, "These companies and that hospital have lost their way."

This is not a moral judgment about any particular act or omission, but about the nature of these companies—their corporate character. Marriage counselors say that a marriage has moved significantly closer to disintegration when one spouse criticizes the other, not for some particular act or omission, but for being a particular kind of person—from "You forgot to take out the garbage" to "You are a lazy SOB." The judgment of a person's character signals a change in the way we are looking at a person—not as someone who just made a mistake, but as someone who makes mistakes, not as someone who failed to do something, but as the kind of person who fails to do what needs to be done, not as someone who lied, but as a liar. Once that move is made and a person's character is put into question, everything the person does is open to question. Where we once presumed a good character, we now presume a bad one, and where we once assumed a problem was a mistake, out of character for the person, we now assume it is exactly what that sort of person would do.

The judgment of character is a judgment about the internal morality of the person, about the kind of person they are, and that is the sort of judgment we are making about Baxter, Guidant, and the hospital: they did not just make a mistake, but responded in a way that illustrated their true corporate character. They were more concerned about maximizing their profits than about doing what they are obligated to do because of the sorts of companies they are. They are like toy companies that purvey toys contaminated by lead and seem unable to ensure that their toys are lead-free. Parents have no way of determining before purchasing a toy whether it contains lead or not, and after purchase, it would be an undue burden on them to have a toy tested for lead before giving it to a child. We rightly expect toy companies to bear that burden and ensure that the toys they sell are safe. Just so, we rightly expect companies like Baxter to do what they can to ensure that their products are safe and are safely dispensed—especially when they have the sort of warning Baxter received from the deaths of three children in Indianapolis. Its failure to answer that wake-up call should create doubt on the part of consumers about the company's commitment to their well-being.

The lesson for engineers, and, indeed, for any professional, is that they display their professional character, the inner morality of the position they have come to occupy in being a professional of a certain sort, in a variety of ways—in what problems they choose, by how they solve those problems, and by how they respond to the inevitable mistakes they make and the failures that occur. We are not interested in their personal morality, to emphasize—what they think about abortion, for instance, or war and peace—but in the inner morality they have as engineers, the way of looking at the world and of solving problems they come to learn as they learn to be engineers.

Engineers whose design solutions have high failure rates under the stress of ordinary use will display their professional character, good or bad, in their responses. Figuring out why the failure rate is so high and redesigning the artifact to reduce the rate to an acceptable level is what engineers are to do—not what great engineers are to do, or even good engineers, but even minimally competent engineers. It is part of the job description of an engineer, that is, that an engineer does not leave failures alone, but goes back to fix them. We would laud praise upon the engineer were the problem solved with a minimum of fuss and expense. We would give extra credit for elegant solutions. We would be reluctant to use an engineer on another project who said, "People must be using it wrong. That's not my problem." If a design provokes errors, it is a problem for the engineer—as well as for those who have been provoked into a mistake because of the error-provocative design.

We may make this clearer by looking at a particular example, an engineer who is assigned the task of designing a better land mine.[101] Some may fault the engineer morally for agreeing to such a task. The only use to which a land mine can be put is to kill and maim those who unknowingly step on or near it, and it is indiscriminate in whom it kills, taking innocent lives of children as well as others. So a moral question must be raised about whether an engineer has a moral responsibility not to work on such a project. But that moral question is an external moral question, a question about what an engineer ought or ought not to do *as a citizen*. The question we are concerned with is what an engineer *as an engineer* ought or ought not to do *given* that the engineer has the task of designing a better land mine.

Suppose the engineer designed a land mine that exploded if it was jostled in any way in shipping? Or that exploded when moved off the horizontal axis even slightly by anyone placing it in a hole? Or that tended to explode while being manufactured? Or some of whose parts so easily corroded that it exploded unexpectedly—or not at all? Or exploded after being stored for awhile—and so set off the other explosives being stored? Or whose failure rate was over 80%? Or whose explosive charge often sputtered instead of exploding?

It might be that the engineer had decided to sabotage the land mine, making those manufacturing and using the land mine pay a steep price for taking part in something so heinous or ensuring that the mine would not harm anyone because it would fail to work. We would then say that the engineer had decided that, *as a citizen*, the best protest was to undermine the land mine, taking it on as a project only to ensure that it would not be done properly.

If the aim is not sabotage, however, and the engineer designed a land mine with any of those faults, we would fault the engineer for failing to fulfill the minimal standards we expect any engineer to uphold. "Not much of an engineer," we would mutter, and we would assign the task to a competent engineer and fire the one who claimed to be an engineer. If that person screwed up the land mine design so badly, what could you trust the person to do on any other project?

We should have the same sort of response to such an engineer as we should have to those in charge of a drug company that fails the patients who rely on it, to those in charge of a heart implant company that puts those who uses its implants at greater risk, to those in charge of a hospital seemingly more concerned about its financial well-being than its patients'. We can no longer be sure that they are doing properly what we give them a license to do. We are making a judgment about the corporate characters of the companies and about the engineer's professional character.

Engineers enter into a form of life that requires them to learn to think in a way very different from the way, say, lawyers or physicians think, and, as we have argued, that form of life has its requirements—a set of skills that must be mastered, certain knowledge that must be obtained, and a level of competence in using those skills and mastering that knowledge. These are requirements that are essential to, built into, living that form of life. A person cannot be an engineer without those skills and that knowledge. But they make their own demands.

To have a particular skill is to know how to act under certain circumstances. Failing to act in the proper way, given those circumstances, may be explained away as a fluke, but it will not take much more than one instance of failure for us to deny the person has the skill. A red flag is raised by a surgeon who fails to recognize a swollen appendix and so fails to remove it, or who sees it, but fails to remove it properly, or who removes it properly, but fails to remove it from the patient before stitching the patient up. We can think of a myriad of reasons for such a failure. But if it happens more than once, we will be loathe to be the surgeon's patient and presume that those who became the surgeon's patients were unaware of these problems. The surgeon's behavior in those circumstances is more than we need to deny that the surgeon has the appropriate skills to be operating.

Just so for someone who claims to be an engineer but fails to exhibit the set of skills and kind of knowledge incumbent upon someone who has entered into that form of life. Imagine an engineer who keeps making mistakes in calculating, failing to take the time or have the patience to do it properly. The skill of calculating may seem easy, but it is fraught with importance and pregnant with potential mistakes. A mistake can enter in a variety of different ways, from punching in the wrong data, to reading the conclusion wrongly, to having a program in place that will not guarantee the right answer, and, as we saw with the Mars lander, a mistake can be costly. So we will ultimately banish from the ranks of engineers someone who keeps making mistakes in calculating. That person lacks a requisite skill.

I refer to the skills and knowledge of a profession as its inner morality because these function as norms. We criticize someone who claims to be a lawyer, but fails to know how to make out a will or fails to exercise due care in making one out so that the will is invalid. Just so, we criticize someone who claims to be an engineer, but fails to use properly the skills necessary to be an engineer or marshal the knowledge needed. Such a person has failed to fulfill the inner morality of the profession. Someone who cannot calculate carefully is no engineer.

9

Engineering and ethics

It is an unfortunate, but standard understanding of the relation between engineering and ethics that the two differ fundamentally, the former requiring the hard numbers of science and math, the latter appealing to something subjective.

The concern is that if ethics were integral to engineering practice, it would be worse off, its quantitative purity muddied by qualitative matters.

But as we have seen, engineers already engage in ethical practice in solving design problems. They cannot help but do that since any proposed solution reflects a choice of values and when realized in an artifact will have effects, some good, some harmful. Ethical considerations permeate design solutions and thus the intellectual core of engineering.

But that engagement needs to be made explicit to change both how engineering is taught and how engineers conceive of themselves and their role in solving design problems.

Ethics internal to engineering

In the Preface to *Essentials of Engineering Design*, to cite a typical remark, Joseph Walton says that the last of the ten chapters "raises ethical questions that an engineer may face from time to time, the non-mathematical problems that need more than a calculator to answer."[102] This remark is typical in two ways:

(a) It implies that engineers will run into ethical problems only occasionally and thus that ethics is not essential to engineering. The issue can be put in the last chapter because it will not make much difference if the class does not get to it since only from "time to time" will the students face any ethical questions as engineers.

(b) It implies that ethics and engineering differ fundamentally. Engineers pose the sort of problem solved by using calculators while ethics poses "the non-mathematical problems" for which calculators are useless. Engineering is quantitative; ethics is not. The implication is that if ethics were integral to engineering practice, engineering practice would be worse off, its quantitative purity muddied by qualitative matters.

This story of the relationship between engineering and ethics is a popular one, the contrast upon which it depends permeating our understanding of how the arts and the sciences differ from one another. It is no wonder, given such an understanding, that engineers may blanch at the idea that ethics is integral to engineering. The narrative of engineering is that there are right answers to engineering problems, answers determined by the nature of the problem and not by what anyone may wish or hope or prefer. Either a particular metal will survive the stresses when used to fabricate the girders of a bridge or it will not. If it does not, the girders and, presumably, the bridge will not survive. So engineers must calculate—using a calculator—the right answer to questions about metals and stress. If ethics were integral to engineering, by this narrative, the right answers risk being overwhelmed by issues about which there are no right answers, issues determined not by any calculations, not determined, in fact, at all, but subject to the vagaries of subjective bias and preference. This understanding of the nature of engineering and ethics is held by engineers and others alike. But, as we now know, it is mistaken.

Ethics permeates design solutions. We have concentrated upon design problems and upon the way in which ethical considerations enter into design solutions because solving design problems is the intellectual core of engineering. You cannot be an engineer without solving design problems, and so if ethical considerations enter into any solution, you cannot be an engineer without taking on a responsibility to be ethical—whether you recognize you have that responsibility or not.

As we saw with the example of designing a pick to get food and other such things from between one's teeth, no design problem determines any one solution. There is thus no single "right answer" to a design problem. There is space for creativity and innovation, with a myriad of design solutions possible for any single design problem—as the various kinds of toothpicks illustrate. There are right answers, of course, to some issues that arise because of a design problem. Some wood will not do for toothpicks, for example, because the stress upon a toothpick when used to pick teeth is too much, and some material will not do because the toothpicks are too rigid and damage a tooth's enamel. These matters are quantifiable. But quantitative considerations alone do not determine a design solution. An engineer's decision about what to do to solve a particular design problem does not rest wholly on the crystalline clarity that quantification supposedly provides, but on ethical considerations.

We may perhaps see this more clearly by thinking about what happens when a design solution is embodied in an artifact—with no misstep between the solution and the artifact. On the one hand, we can readily imagine an evil genius of an engineer, and on the other hand, we can imagine an engineer adopting the primary principle that the solutions be benign by design. We can readily imagine, that is, the worst and the most benign of design principles: designing so as to ensure that harm will result and designing so as to ensure, as best we can, that no unnecessary harm will result. And so we can readily imagine the artifacts that would result from the design solutions of these different engineers. They will have different causal effects when introduced into the world, the one by the evil genius of an engineer producing harms because it was designed to do so, the one by the benign engineer minimizing harms because it was designed to minimize them.

We need not invoke an engineer's intentions, however, to show how ethics enters into the intellectual core of engineering. Because an engineer's design solutions themselves embody differing sets of values, and because the artifacts that embody those solutions have different sets of effects, with different configurations of harms, an engineer is responsible for whatever design solution is chosen. What I have called the arguments from design and effects make this point. An engineer is thus responsible for whatever unnecessary harms come from the artifact that fully realizes that design solution. Like all of us, the engineer is morally obligated not to cause unnecessary harm.

When an engineer does not fulfill that obligation, we can end up with an accident—like the crash of the Colombia airliner with all its attendant harms, 159 people dead and a plane destroyed with the added expense of having to check all autopilot software, retrain pilots, pay those who sue, and on and on. The circumstances are not responsible for the harms. The pilot is not responsible for the harm. The software engineers are. If they had been evil geniuses, we would fault them morally for intending such terrible harms.

But even without any evil intent, we should hold them morally accountable. That they did not intend to cause such harms is morally irrelevant. Competent engineers should not produce such shoddy work, and an engineer who does is properly held morally accountable for incompetence.

Any time we introduce harm or what could cause harm into the world, we have a moral problem if we are in a position to preclude that possibility—if the harm is gratuitous because it need not be introduced. That we can imagine an evil genius of an engineer and a benign engineer is all the proof needed that ethics is integral to the design process. Engineering artifacts, that is, can be designed to cause great harm and to be as benign as possible. The latter is morally preferable because it is morally wrong to cause gratuitous harm. In solving the design problems that are the intellectual core of engineering, to repeat, engineers thus have a moral obligation to provide solutions, as best they can, that at the least do not cause unnecessary harm. That means that the engineering decision that such-and-such is the best solution to a design problem has moral weight—a negative weight when it causes harm that some other solution, equally acceptable, would not, a positive weight if it is at least benign.

So ethics enters into the intellectual core of engineering because the design solutions that engineers make themselves embody moral values and because an artifact that embodies that design decision will have its effects in the world, including more or less harm. If the design solution, once realized in an artifact, will cause more harm than necessary, that solution is morally a mistake. Engineers cannot engage in the core enterprise of engineering, solving design problems, without making moral choices in solving the problems with moral implications once those solutions are embodied in artifacts. A consideration of how ethics enters engineering deserves to be in the first chapter of any engineering text, not the last.

Engineers also cannot engage in that core enterprise without making use of the knowledge and skills they must learn in order to become engineers, and, as we saw, they have a moral obligation to make use of that knowledge and to make, as it were, skillful use of their skills. Just as the design solution proposes a rule for how to solve the design problem—"If you want to have a bridge that reaches from this side to that and is strong enough to support heavy trucks, then here is what you need to do"—so creating a design solution requires the use of rules about how to measure, about what sorts of materials are appropriate in what kinds of situations, and on and on. Engineers can go wrong—morally wrong—by failing to use the proper rules of skill or, using them, failing to use them properly, and they can go wrong by not being prepared with any clear rule at all when they ought to have a vetted rule in hand, prepared ahead of time for such possible catastrophic eventualities.

Engineers take on the moral responsibility to use those knowledge and skills properly in becoming engineers. That responsibility is a feature of the moral relations that attach to that role. As we have seen, an engineer is no different from any other professional in this regard. A lawyer must take due care in making a will, ensuring that it is properly filled out, notarized, and registered. A failure in any one of these regards will invalidate a will, causing problems to those who were to inherit, among others. A broker selling equity in a company must exercise due diligence in investigating that company, ensuring that those making purchases from the broker have what they need to make an informed decision. And engineers need to assess the strength of the materials to be used in building a bridge, for instance, to ensure that the material chosen is strong enough to withstand the stresses the bridge will undergo, double checking the measurements of the solution they propose to a design problem to ensure that everything will work together as envisaged, and so on.

A full understanding of how ethics enters engineering would need to consider how morality affects what we see as a problem and how we brainstorm to solve problems. Engineers are trained to see certain kinds of problems—just as are all professionals. An engineer may look at an apple tree and see nothing but good-looking apples almost ready to pick. A horticulturist may look and see a tree that needs serious help. An engineer may be working with team members in tension without realizing what a psychologist would spot right away.

The story about the engineer about to be guillotined makes the point well. Presumably, the engineer saw that he was about to be guillotined, but could not help himself from thinking like an engineer and so looking up to see if he could find the problem. He then could not help himself from acting like an engineer: "There's a problem here that needs to be fixed." A Wall Street trader who saw the problem might say, "Bet you a million to one God will grant me a miracle too!" Not the engineer.

Exploring how values affect what engineers, or any professionals, see as problems opens up a universe of issues we shall not examine in any detail here. Some of these issues arise because of social factors that are not likely to register for those working within the society. An American lawyer sees a dispute as a first step to its final resolution before a judge. A Japanese lawyer sees a dispute as a problem that needs to be resolved so that it does not go before a judge where one party will win and both parties will thus lose face since even the winner will lose face, given the cultural values, by making someone else lose face. An American lawyer's job is thus to sharpen the points of contention; a Japanese lawyer's job is to find the areas of compromise so that the dispute is resolved in a way that satisfies both parties. These differing modes of thought reflect differing social values. Japanese lawyers lose face, and business, if their clients have to go to court to settle a dispute; American lawyers revel in a court victory.

Besides the professional and social differences, there are no doubt personal differences too between different professionals within one society and within a profession. One engineer may look at a car and wonder how to make it go faster; another may look and wonder how to make it greener. We will not pursue here the value implications of the initial sighting of a problem by engineers, but we can understand how rich a source of examples, and of lessons to be learned, we could find by a full examination. The problems they see as well as the solutions they offer are not morally neutral.

We will also not pursue here how morality affects brain-storming, affects, among other things, the range of possible solutions an engineer facing a problem engenders. We are witnesses now to a sea change in the scope of possible solutions in the green revolution we are in. Light bulbs had remained essentially unchanged since they were invented, and it is only now, in the green revolution, that engineers have seen their present configuration as a problem because incandescent bulbs waste too much energy. It is only now that a change in the moral atmosphere, as it were, has triggered the brain storms that are transforming how we light up our world.

In any event, my aim has been to show that ethics is integral to engineering. That it is integral to design solutions more than suffices to make that point. Ethics in fact permeates engineering practice—from the perception of a problem as an engineering problem to the end life of the artifacts that embody design solutions. But each point at which ethics enters requires different considerations and different arguments.

Four responses

1. **"We are not responsible for everything."**—One of the standard responses to the question of legal liability, and no doubt moral liability as well, is that engineers cannot be held responsible for all the things people do with what they create. If someone drives a car recklessly, the driver is at fault, not the

engineers. That is surely correct: engineers are not responsible for everything anyone does with the artifacts that embody their design solutions. There are too many idiots in the world, of too many different kinds, for engineers to foresee and so forestall the mistakes people will make, let alone the silly things they will do with engineering artifacts because they are just not thinking or thinking well. Using a screwdriver to test whether the electric line coming in from outside is hot is not something any engineer is likely to design a screwdriver to survive.

The Darwin awards provide us with a slew of examples, but this one about the screwdriver is drawn from real life. One year when I was away and rented our house, a tenant used one of my screwdrivers for that purpose. I discovered this when I went to get a screwdriver and found the one in question neatly put away, with a blackened handle and almost no blade. I asked him if he knew what had happened to the screwdriver. He said that the lights had gone out one time and that he was not sure whether the problem was internal to the house or general. "So I stuck the screwdriver where the big wire comes in from outside." He added, "Boy, was I surprised! It just flew out of my hand. I had to dig it out of the wall over there." Engineers are not responsible for *everything* people do with engineering artifacts.

They are responsible, however, for some things people do with those artifacts. That is one point of the example of error-provocative designs. If the artifacts that realize the engineers' design solutions provoke errors for those who use them, the engineers are morally responsible. If the design is so bad that even the most intelligent, well-trained, and highly motivated operator is provoked by the design into making an error that causes harm—as with the software in that Colombia airliner autopilot—then it would be disingenuous in the extreme for an engineer to say, "I'm not responsible for what people do with what I design!" All we need to do is to imagine an evil genius of an engineer who purposefully creates such designs and then accuses those who use the artifacts realizing those designs of causing harm. Not being responsible for everything people do with engineering artifacts does not get engineers off the moral hook of responsibility for some of what people do with engineering artifacts when what they do is triggered by a faulty design.

Besides, responding that engineers are not responsible for what people do with engineering artifacts presupposes that the only moral issue that arises in engineering concerns the use by an operator of such artifacts, but, as we have seen, morality enters in other ways. Designing a switch that uses mercury when alternative design solutions were feasible is morally wrong not because anyone is going to misuse the switch, but because of the problems of disposing of the mercury in the switches without harm to us and to our environment. Designing an airbag that disadvantages women and small children and advantages males is morally wrong not because anyone does anything with the airbag, but because some drivers had no choice but be put in a far riskier position than they would have been without the airbag.

2. **"We could design something completely safe."**—That brings us to the second response my view is likely to provoke. It is sometimes said that engineers could design something that

was completely safe to drive—a tank rather than a car, I have heard it suggested—but that no one would be able to afford it or drive it because it would be so heavy and so well-armored. That may be true, but is irrelevant. Safety should not be all that engineers ought to be concerned about, and ensuring that individuals are safe is not all there is to ethics. When our software reports "Unknown Error 0x80040119," when the trunk of our Cadillac breaks because someone closed it the way we normally close trunks, when we cannot wash our hands because we cannot get a faucet to work, and on and on, no one is harmed by these problems in the sense of having some physical injury, but our interests are set back, and so we are harmed in that way. Safety is not at issue regarding such engineering artifacts, and that makes the point that safety is not the only concern engineers have in making their design solutions moral.

3. **"And, besides, we are already ethical!"**—The third response is a form of disbelief. I can hear engineers telling me, "We're not evil! We're already moral! You are just describing what we already do!" Well, sort of. I am certainly not suggesting that the engineers are unethical. Quite the contrary. We would live in a far worse world if they were unethical, since their fingerprints are all over our technological universe, and if engineers were generally perversely evil, they could wreak havoc for us. So engineers generally do what they ought to do. I am thus describing what is already the general practice of engineers at least in regard to some ethical matters—e.g., safety. But I am suggesting that once engineers realize that

their current practice is driven by moral considerations, they will widen the scope of their concerns about minimizing potential harms and self-consciously strive to produce morally better design solutions. They will consider more conscientiously, for instance, the life cycles of the artifacts that realize their design solutions and solve design problems in ways that will at the minimum minimize harms.

4. **"And if we are not already ethical, where are we to find the ethics in this book?"**—Engineers may well wonder, "Where's the ethics? Where's utilitarianism? Where's virtue theory? Where's Kant? How do I resolve ethical problems? How do I know I've got an ethical problem?" There is a hint of virtue theory in our discussion of role morality: we take on certain features, i.e., virtues, when we come to occupy a role, and we will need to develop those features to become the best at them that we can be. But there is no discussion of the big ethical theories, no attempt to show how they relate, if at all, to engineering practice, no discussion of how they might help resolve ethical problems engineers face. There are reasons for that.

If engineers adopt the principle "Benign by design!" and conscientiously consider how to avoid harms, they will avoid many ethical problems that may arise because of design choices they would otherwise have made. The best protection against having to settle moral problems, by appeal to moral theory, is to prevent their occurrence, and the best way to prevent their occurrence in engineering is to avoid unnecessary harms—of all sorts. Engineers do not need any ethical theory to justify avoiding unnecessary harms.

Indeed, putting ethical theories between an engineer and understanding that ethics is integral to engineering is both unhelpful and harmful. We do not need any ethical theory to know that we should not cause unnecessary harm, and it is that ethical principle which permeates the text. Each ethical theory justifies that principle in different ways, but the different justifications do not matter for the claim here that ethics is integral to engineering. Each ethical theory is complex and sometimes more than a little difficult to understand, to put it mildly, and if ethical practice depended upon understanding ethical theory, we would have far less ethical practice than we now do. And the leading contenders for the ethical theory are at odds with one another, each holding up a different vision of how we ought to live our lives. Philosophers cannot agree on which is morally preferable, and if philosophers cannot agree, we can hardly expect engineers to digest these theories and make a rational and moral choice between them before they engage in engineering practice. They do not need to.

They will need to weigh and assess competing harms and benefits. Issues may arise about the weight of competing harms, for instance. Is it morally preferable to solve a problem completely now, even with attendant harm, or to minimize the problem rather than eliminate it, but with far less harm? Is it morally preferable to choose a material for an artifact that takes less energy to produce than a material that takes more energy to produce but will break less easily? It is not obvious how to make such decisions, but philosophers are not obviously any better situated than engineers to do that. Engineers are at least used to weighing and assessing all sorts of competing demands for the artifact that realizes their design solution—inexpensive to produce, long shelf life if that is relevant, easy to ship without breakage, and so on and so on—and so taking harms explicitly into account and assessing and weighing them is not far outside their experiential base. They appear far better positioned than any philosopher, that is, to make such assessments once it is made clear it is harms and benefits that are being weighed and assessed and that they have, at the minimum, a moral obligation to minimize potential harms.

These responses will probably not, and should not, exhaust the list of concerns engineers will have about the claim that ethics is integral to engineering practice. If it is integral, and engineers recognize that it is integral, then that practice is going to change and how we teach engineering must change. In teaching students how to think like an engineer, we cannot just focus on the quantitative features that are, indeed, essential to good engineering practice, for instance. We are going to have to emphasize imaginative and creative thinking, a working understanding of how we think about and so approach the artifacts of our lives, and a sense of the history of a design solution. We do not want the artifact that realizes a new design solution to stymy us because of the habits we bring to it, and we should take full advantage of the creativity of previous engineers who have thought about the issues and perhaps come up with designs that need to be revived. I have focused on how ethics enters engineering for a lone engineer trying to solve a design problem, but engineering is a social enterprise, with a long history of success and of failure, and engineering education needs to reflect that history if engineers are not to rely just on their own creative resources.

NOTES

Foreword

1   Hard or soft (or professional) rules are what Kant called rules of skill, and drawing that sort of distinction, hard and soft, between the two is misleading. The distinction makes it sound as though there are difficult rules to learn, the hard ones, and easy ones, the soft ones. But it can surely be as hard to communicate clearly as it can be to calculate stresses. Even a master of communication like Edward Tufte can fail the test of clear and perspicuous communication because, as it turns out, it is not easy to make things clear. See my "Representation and Misrepresentation: Tufte and the Morton-Thikol Engineers on the Challenger," with Roger Boisjoly, David Hoeker, and Stefan Young (*Science and Engineering Ethics* 8, 2002, pp. 1–31) and my "Ethical Presentations of Data: Tufte and the Morton-Thiokol Engineers," in *Philosophy and Engineering: Exploring Boundaries, Expanding Connections*, eds. Diane P. Michelfelder, Byron Newberry, Qin Zhu (Dordrecht: Springer, forthcoming).

2   See in this regard Richard K. Miller, "Why the Hard Science of Engineering Is No Longer Enough to Meet the 21st Century Challenges," esp. pp. 8ff. Online at http://www.olin.edu/sites/default/files/rebalancing_engineering_education_may_15.pdf (accessed December 7, 2015).

3   http://www.abet.org/wp-content/uploads/2015/04/eac-criteria-2013-2014.pdf. I suspect that those who formulated the ABET soft-skill list would be surprised that ethical considerations enter into the design process and are thus, in that way, essential to an engineer's "understanding of professional and ethical responsibility." I suspect they meant instead such ethical requirements as being honest.

Chapter 1

4   Lucy Perkins, July 28th, "Human Error Caused Virgin Galactic Crash, Investigators Say," July 28, 2015, National Public Radio. Online at

http://www.npr.org/sections/thetwo-way/2015/07/28/427160185/human
-error-caused-virgin-galactic-crash-investigators-say. Accessed August 9, 2015.

5   Alex Davies, "Blame a Catastrophic Blindspot for the Fatal Virgin Galactic
    Crash," July 28, 2015, *Wired*. Online at http://www.wired.com/2015/07/blame
    -catastrophic-blindspot-virgin-galactic-crash/. Accessed August 9, 2015 .

6   Online at http://www.goodexperience.com/tib/archives/product_design//.
    Accessed January 18, 2010.

7   Henry Petroski, *The Toothpick: Technology and Culture* (New York: Alfred
    A. Knopf, 2007), pp. 250–51.

8   Edward Barnett, "Periodontal and Dental Cleanser and Periodontal
    Stimulator," US Patent No. 3,775,848 (December 4, 1973), quoted in
    Petroski, *The Toothpick*, pp. 263–64.

9   Petroski, *The Toothpick*, p. 262. The patent was issued on August 23, 1923,
    to Russell Edward Lunday as Patent No. 1,465,522.

10 Henry Petroski, *The Evolution of Useful Things* (New York: Alfred A. Knopf,
    1995), pp. 22–33.

Chapter 2

11 Micheline Maynard and Matthew L. Wald, "Experts Puzzle over How
    Flight Overshot Airport," *New York Times*, October 23, 2009.

12 Gene Park, "FAA probes whether Go! pilots fell asleep," *Honolulu Star
    Bulletin*, Vol. 13, No. 51, February 20, 2008. Online at http://archives
    .starbulletin.com/2008/02/20/news/story08.html. Accessed July 5, 2015.

13 Tim Hume, "Captain of TransAsia Flight 235 shut off working engine
    after other failed: Report," CNN, July 2, 2015. Online at http://www.cnn
    .com/2015/07/02/asia/taiwan-transasia-crash-report/. Accessed July 6, 2015.

14 "What Went Wrong on the Day the Music Died?," *All Things Considered*,
    National Public Radio, February 2, 2009. Online at http://www
    .npr.org/templates/story/story.php?storyId=100209015. Accessed
    February 27, 2015.

15 Will Stewart, "Russian roulette shock as wedding guest shoots himself in
    party trick gone wrong," *Daily Mail*, March 23, 2010. Online at http://www
    .dailymail.co.uk/news/article-1259841/Russian-roulette-shock-wedding

-guest-shoots-party-trick-gone-wrong.html#ixzz3Nl807ZmU. Accessed
January 3, 2015. See also http://darwinawards.com/darwin/darwin2000-04
.html. Accessed January 3, 2015.

16   http://abcnews.go.com/travel/aheadofthecurve/story?id=5930052.
     Accessed February 27, 2015. See also Jesse McKinley and Matthew
     LWald, "California Bans Texting by Operators of Trains," *New York Times*,
     September 19, 2008.

17   Matthew L.Wald, "Expert Says Engineer Sent Text Messages Before Deadly
     Train Crash," *New York Times*, January 22, 2010.

18   Joe Morgenstern, "The Fifty-Nine-Story Crisis," *The New Yorker*, May 29,
     1995, pp. 45–53.

19   Wade Robison et al., 2002, pp. 1–24.

Chapter 3

20   "Pilot's Wrong Keystroke Led To Crash, Airline Says," *New York Times*,
     August 24, 1996, p. 9.

21   Stephen Manes, "When Trust in 'Data' Is Misplaced," *New York Times*,
     September 17, 1996, p. C9.

22   "Pilot's Wrong Keystroke Led to Crash, Airline Says," *New York Times*,
     August 24, 1996, p. 9.

23   Ibid.

24   Don Phillips, "Putting the Pilot Back in Autopilot," *The Washington Post
     National Weekly Edition*, April 29–May 5, 1996, p. 19.

25   John Markoff, "Report Cites Dangers of Autonomous Weapons," *New York
     Times*, February 28, 2016.

Chapter 4

26   Cited in Martin Helander, *A Guide to Human Factors and Ergonomics*, 2nd
     edn. (New York: Taylor & Francis, 2005), pp. 101–02.

27 Brian Naylor, "Toyota Recall Shines Harsh Light on Safety Agency," Morning Edition, National Public Radio, February 4, 2010.

28 Jim Motavalli, "Runaway Toyotas? Investigation of sudden acceleration eerily recalls deadly Ford transmission issue 25 years ago," Mother Nature Network. Online at http://www.mnn.com/transportation/cars/blogs/ runaway-toyotas-investigation-of-sudden-acceleration-eerily-recalls -deadly-Ford-transmission-issue-25-years-ago. Accessed February 4, 2010.

29 Bill Vlasic, "G.M. Enquiry Cites Years of Neglect over Fatal Defect," New York Times, June 5, 2014. Online at http://www.nytimes.com/2014/06/06 /business/gm-ignition-switch-internal-recall-investigation-report.html. Accessed January 6, 2015.

Chapter 5

30 Mark Mazzetti and Matt Apuzzojan, "C.I.A. Officers Are Cleared in Senate Computer Search," New York Times, January 14, 2015.

31 Stephanie Saul, "Some Sleeping Pill Users Range Far Beyond Bed," New York Times, March 8, 2008; "Ambien in the Driver's Seat," New York Times, March 11, 2008; Stephanie Saul, "Study Links Ambien Use to Unconscious Food Forays," New York Times, March 14, 2008.

32 M'Naghten's Case (1843) 10 C & F 200.

33 Wade Robison, "In the Moral Zone," Teaching Ethics, Vol. 8, No. 2, Spring 2008, pp. 57–78.

34 Rick Bragg, "New Trial Is Sought for Inmate Whose Lawyer Slept in Court," New York Times, January 23, 2001.

35 Linda Greenhouse, "Inmate Whose Lawyer Slept Gets New Trial," New York Times, June 4, 2002.

36 Walt Bogdanich, "At V.A. Hospital, a Rogue Cancer Unit," New York Times, June 21, 2009.

37 Gilbert Ryle, The Concept of Mind (London: Hutchinson, 1949).

38 Aristotle, Nicomachean Ethics, trans. Martin Oswald (Indianapolis: Library of Liberal Arts, 1962), p. 43.

39 Aristotle, Nicomachean, p. 42.

40 Online at http://www.reuters.com/article/idUSTRE60P50O20100126. Accessed February 1, 2010.

41 Online at http://english.pravda.ru/society/stories/06-10-2006/84912 -amputation-0. Accessed February 1, 2010.

42 "Doctors Fined for Fight in Operating Room," *New York Times*, November 28, 1993.

Chapter 6

43 Aristotle, *Nicomachean Ethics*, trans. J. A. K.Thomson in *The Ethics of Aristotle* (London: Penguin, 1953), p. 66.

44 Joel Feinberg, *Harm to Others*, Vol. 1 of *Moral Limits of Criminal Law* (Oxford: Oxford University Press, 1987).

45 Don Gotterbarn, "Computer Ethics: Responsibility Regained," *National Forum: The Phi Kappa Phi Journal*, Vol. LXXI, 1991, p. 2; Terrell Ward Bynum and Simon Rogerson, *Computer Ethics and Professional Responsibility* (Oxford: Blackwell, 2004), p. 110; Robert Riser and Don Gotterbarn, "Ethics Activities in Computer Science Courses." Online at http://csciwww.etsu.edu /gotterbarn/ArtTE2.htm Accessed September 27, 2015.

46 Barry Meier, "Defective Heart Devices Force Some Scary Medical Decisions," *New York Times*, June 20, 2005.

47 Barry Meier, "Repeated Defect in Heart Devices Exposes a History of Problems," *New York Times*, October 20, 2005.

48 Ibid.

49 *Report of the Independent Panel of Guidant Corporation*, March 20, 2006, p. 33.

50 Timothy Williams, "Citing Failures, Guidant Will Recall Thousands of Defibrillators," *New York Times*, June 17, 2005.

51 Williams, "Citing Failures".

52 Barry Meier, "Concern About New Design for Heart Devices," *New York Times*, December 11, 2008.

53 Robert G. Hauser, "A Better Method for Preventing Adverse Clinical Events Caused by Implantable Cardioverter-Defibrillator Lead Fractures?" *Circulation*, Vol. 118, 2008, p. 2117ff.

Chapter 7

54 Jeremy W. Peters, "MOTORING; Unraveling the Mystery of Ford's Fire-Prone Switches," *New York Times*, August 27, 2006.

55 David Barboza, "Firestone Workers Cite Lax Quality Control," *New York Times*, September 15, 2000.

56 Andrew Martin, "Chinese Tires Are Ordered Recalled," *New York Times*, June 26, 2007.

57 James Barron, "The Blackout of 2003: The Overview; Power Surge Blacks Out Northeast, Hitting Cities in 8 States and Canada; Midday Shutdowns Disrupt Millions," *New York Times*, August 15, 2003. See also http://en.wikipedia.org/wiki/Northeast_Blackout_of_2003 Accessed April 17, 2014 .

58 See http://wikicars.org/en/Airbag. Accessed April 17, 2014.

59 Carin M. Olson, Peter Cummings and Frederick P. Rivara, "Association of First and Second-Generation Air Bags with Front Occupant Death in Car Crashes: A Matched Cohort Study," *American Journal of Epidemiology* 164:2, p. 161.

60 Roger A. Saul, Howard B. Pritz, Joeseph McFadden, Stanley H. Backaitis, Heather Hallenbeck and Dan Rhule, "Description and Performance of the Hybrid III Three-Year-Old, Six-Year-Old and Small Female Test Dummies in Restraint System and Out-Of-Position Air Bag Environments," Transportation Research Center Inc. United States Paper Number 98S7-O-01.

61 http://www.sfchronicle.us/cgi-bin/article.cgi?f=/n/a/2009/08/10/national/w160656D02.DTL&type=science#ixzz0OXFDGgXX. Accessed April 23, 2014.

62 Wade Robison, *Decisions in Doubt: The Environment and Public Policy* (Hanover, New Hampshire: University Press of New England, 1994).

63 Daimler, Sustainability Report, 2011. Online at http://sustainability.daimler.com/reports/daimler/annual/2012/nb/English/4545/recycling.html. Accessed January 21, 2015.

64 Askiner Gungor and Surendra M. Gupta, "Issues in Environmentally Conscious Manufacturing and Product Recovery: A Survey," *Computers & Industrial Engineering* 36, 1999, p. 823. See also S.B. Billatos and V.V. Nevrekar, "Challenges and Practical Solutions to Designing for the Environment," ASME Design for Manufacturability Conference, Chicago, IL, March 14–17, 1994, pp. 49–64.

65 John Voelcker, "Who Knew? A Car Battery Is the World's Most Recycled Product," *Green Car Reports*, March 31, 2011. Online at http://www .greencarreports.com/news/1044372_who-knew-a-car-battery-is-the -worlds-most-recycled-product. Accessed January 21, 2015.

66 Joel Feinberg, *Harm to Others*, Vol. 1 of *Moral Limits of Criminal Law* (Oxford: Oxford University Press, 1987).

67 Henry Petroski, *To Engineer Is Human: The Role of Failure in Successful Design* (New York: Vantage Books, 1992), pp. 85–93.

68 See http://www.boston.com/news/specials/big_dig_ceiling_collapse/. Accessed August 27, 2015, for a list of articles from the *Boston Globe* chronicling the various problems with the Boston tunnel and, in particular, the problems with the anchoring system for the ceiling tiles, one of which fell and crushed a woman in a car.

69 Matthew L. Wald and Kenneth Chang, "Minneapolis Bridge Had Passed Inspection," *New York Times*, August 3, 2007.

70 Atul Gawande, "The Way We Age Now," *The New Yorker*, April 30, 2007. Online at http://www.newyorker.com/reporting/2007/04/30/070430fa _fact_gawande?currentPage=2

71 http://clearviewhwy.com/WhatIsClearviewHwy/index.php. Accessed May 28, 2014.

72 http://clearviewhwy.com/WhatIsClearviewHwy/HowItWorks/ letterformDesign.php. Accessed May 28, 2014.

73 http://clearviewhwy.com/ResearchAndDesign/. Accessed May 28, 2014.

74 Joshua Yaffa, "The Road to Clarity," *New York Times Magazine*, August 12, 2007.

75 Yaffa, "The Road to Clarity".

76 http://clearviewhwy.com/ResearchAndDesign/legibilityStudies.php. Accessed June 2, 2014.

77 Yaffa, "The Road to Clarity".

78 https://www.transportation.gov/fastlane/fonts-and-highway-safety. Accessed January 26, 2016.

79 Henry Petroski, "Easy Reading Road Signs Head to the Offramp", *New York Times*, February 26, 2016.

Chapter 8

80 "Doctor Who Cut Off Wrong Leg Is Defended by Colleagues," *New York Times*, September 17, 1995.

81 Paul Sisson, "Kaiser Hospital Fined for Removing Wrong Kidney," *U-T San Diego News*, December 20, 2012. Online at http://www.utsandiego.com /news/2012/dec/20/kaiser-hospital-fined-for-removing-wrong-kidney/. Accessed January6, 2013.

82 "Florida Doctor Who Took Out Wrong Organ Fined $5000," *Insurance Journal*, June 9, 2010; online at http://www.insurancejournal.com/news /southeast/2010/06/09/110568.htm. Accessed January 3, 2013.

83 Linda Greenhouse, "Inmate Whose Lawyers Slept Gets New Trial," *New York Times*, June 4, 2002.

84 Jon Nordheimer, "New Jersey Autopsy Misses Two Bullets in a Man's Head," *New York Times*, October 20, 1993.

85 Aristotle, *Nichomachean Ethics*, trans. Martin Oswald (Indianapolis: Library of Liberal Arts, 1962 p. 43.

86 Online at http://www.typewritermuseum.org/collection/index .php3?machine=blick1&cat=ks.

87 Henry Petroski, *To Engineer Is Human: The Role of Failurein Successful Design* (New York: Vantage Books, 1992), p. 87; see also Matthys Levy and Mario Salvadori, *Why Buildings Fall Down: How Structures Fail* (New York: W. W. Norton & Co., 1994), pp. 221–30.

88 Petroski, *To Engineer Is Human*, p. 89.

89 Ibid., p. 88.

90 Ibid., pp. 86–87.

91 Ibid., p. 89.

92 Todd Mytkowicz, Amer Diwan, Matthias Hauswirth and Peter F. Sweeney, "Producing Wrong Data Without Doing Anything Obviously Wrong!", Fourteenth International Conference on Architectural Support for Programming Languages and Operating Systems, March 7–11, 2009, Washington, D.C.

93 Douglas Isbell, Mary Hardin and Joan Underwood, NASA release 99–113, September 30, 1999.

94 Henry Petroski, *To Engineer Is Human*, p. 164; Levy and Salvadori, *Why Buildings Fall Down*, pp. 107–20.

95 Diane Vaughan, *The Challenger Launch Decision: Risky Technology, Culture, and Deviance at NASA* (Chicago: University of Chicago Press, 1997), pp. 278ff.; Wade Robison et al., pp. 1–24.

96 Vaughan, *The Challenger Launch Decision*, p. 317.

97 Ibid., p. 318.

98 Michael Davis, *Thinking Like An Engineer: Studies in the Ethics of a Profession* (New York: Oxford University Press, 1998), p. 67.

99 Presidential Commission Report, July 6, 1986, NASA, Chapter V. Available online at http://history.nasa.gov/rogersrep/genindex.htm. Accessed May 26, 2015.

100 Tara Parker-Hope, "A Hollywood Family Takes on Medical Mistakes," *New York Times*, March 17, 2008. Online at http://well.blogs.nytimes.com/2008/03/17/a-hollywood-family-takes-on-medical-mistakes/. Accessed December 27, 2015; quotations are from 60 Minutes, March 16, 2008, available at the link to Parker-Hope.

101 Wade Robison, "Design Problems and Ethics," Ibo van de Poel and David E. Goldberg (eds.), *Philosophy and Engineering: An Emerging Agenda*, Vol. 2 of the Series Philosophy of Engineering and Technology (Heidelberg: Springer, 2009), pp. 205–14.

Chapter 9

102 Joseph W. Walton, *Essentials of Engineering Design* (St. Paul: West Publishing Co., 1991), p. xv.

BIBLIOGRAPHY

Aristotle. *Nicomachean Ethics*. Translated by J. A. K. Thomson in *The Ethics of Aristotle*. London: Penguin, 1953.

Aristotle. *Nicomachean Ethics*. Translated by Martin Oswald. Indianapolis: Library of Liberal Arts, 1962.

Barboza, David. "Firestone Workers Cite Lax Quality Control," *New York Times*, September 15, 2000.

Barnett, Edward. "Periodontal and Dental Cleanser and Periodontal Stimulator," U.S. Patent No. 3,775,848 (December 4, 1973).

Barron, James. "The Blackout of 2003: The Overview; Power Surge Blacks Out Northeast, Hitting Cities in 8 States and Canada; Midday Shutdowns Disrupt Millions," *New York Times*, August 15, 2003.

Billatos, S. B. and Nevrekar, V. V. "Challenges and Practical Solutions to Designing for the Environment," ASME Design for Manufacturability Conference, Chicago, IL, March 14–17, 1994, 49–64.

Block, Melissa. "What Went Wrong On The Day The Music Died?," *All Things Considered*, National Public Radio, February 2, 2009.

Bogdanich, Walt. "At V.A. Hospital, a Rogue Cancer Unit," *New York Times*, June 21, 2009.

Bragg, Rick. "New Trial Is Sought for Inmate Whose Lawyer Slept in Court," *New York Times*, January 23, 2001.

Bynum, Terrell Ward and Rogerson, Simon. *Computer Ethics and Professional Responsibility*. Oxford: Blackwell, 2004.

Daimler Sustainability Report, 2011.

Davies, Alex. "Blame a Catastrophic Blindspot for the Fatal Virgin Galactic Crash," *Wired*, July 28, 2015.

Davis, Michael. *Thinking Like an Engineer: Studies in the Ethics of a Profession*. New York: Oxford University Press, 1998.

"Doctor Who Cut Off Wrong Leg Is Defended by Colleagues," *New York Times*, September 17, 1995.

"Doctors Fined For Fight in Operating Room," *New York Times*, November 28, 1993.

Feinberg, Joel. *Harm to Others*, Vol. 1 of *Moral Limits of Criminal Law*. Oxford: Oxford University Press, 1987.

"Florida Doctor Who Took Out Wrong Organ Fined $5000," *Insurance Journal*, June 9, 2010.

Gawande, Atul. "The Way We Age Now," *The New Yorker*, April 30, 2007.

Gotterbarn, Don. "Computer Ethics: Responsibility Regained," *National Forum: The Phi Kappa Phi Journal* LXXI.2 (1991), 26–31.

Gotterbarn, Don and Riser, Robert. "Ethics Activities in Computer Science Courses," http://csciwww.etsu.edu/gotterbarn/ArtTE2.htm. Accessed September 17, 2015.

Greenhouse, Linda. "Inmate Whose Lawyer Slept Gets New Trial," *New York Times*, June 4, 2002.

Gungor, Askiner and Gupta, Surendra M. "Issues in Environmentally Conscious Manufacturing and Product Recovery: A Survey," *Computers & Industrial Engineering* 36 (1999), 823.

Hauser, Robert G. "A Better Method for Preventing Adverse Clinical Events Caused by Implantable Cardioverter-Defibrillator Lead Fractures?" *Circulation* 118.21 (2008), 2117ff.

Helander, Martin. *A Guide to Human Factors and Ergonomics*, 2nd edn. New York: Taylor & Francis, 2005.

Hume, Tim. "Captain of Trans Asia Flight 235 Shut Off Working Engine after Other Failed: Report," CNN, July 2, 2015.

Isbell, Douglas, Hardin, Mary and Underwood, Joan. "Mars Climate Orbiter Team Finds Likely Cause of Loss," NASA Release 99–113, September 30, 1999.

Levy, Matthys and Salvadori, Mario. *Why Buildings Fall Down: How Structures Fail*. New York: W. W. Norton & Co., 1994.

Manes, Stephen. "When Trust in 'Data' Is Misplaced," *New York Times*, September 17, 1996.

Martin, Andrew. "Chinese Tires Are Ordered Recalled," *New York Times*, June 26, 2007.

Maynard, Micheline and Wald, Matthew L. "Experts Puzzle Over How Flight Overshot Airport," *New York Times*, October 23, 2009.

Mazzetti, Mark and Apuzzojan, Matt. "C.I.A. Officers Are Cleared in Senate Computer Search," *New York Times*, January 14, 2015.

Meier, Barry. "Defective Heart Devices Force Some Scary Medical Decisions," *New York Times*, June 20, 2005.

Meier, Barry. "Repeated Defect in Heart Devices Exposes a History of Problems," *New York Times*, October 20, 2005.

Meier, Barry. "Concern About New Design for Heart Devices," *New York Times*, December 11, 2008.

Miller, Richard K. "Why the Hard Science of Engineering Is No Longer Enough to Meet the 21st Century Challenges," 8ff.

Morgenstern, Joe. "The Fifty-Nine-Story Crisis," *The New Yorker*, May 29, 1995.

Motavalli, Jim. "Runaway Toyotas? Investigation of sudden acceleration eerily recalls deadly Ford transmission issue 25 years ago," Mother Nature Network

at http://www.mnn.com/transportation/cars/blogs/runaway
-toyotas-investigation-of-sudden-acceleration-eerily-recalls-deadly.
Accessed February 4, 2010.

Mytkowicz, Todd, Diwan, Amer, Hauswirth, Matthias and Sweeney,
Peter F. "Producing Wrong Data Without Doing Anything Obviously
Wrong!" Fourteenth International Conference on Architectural
Support for Programming Languages and Operating Systems,
March 7–11, 2009.

Naylor, Brian. "Toyota Recall Shines Harsh Light on Safety Agency," Modern
Edition, National Public Radio, February 4, 2010.

Nordheimer, Jon. "New Jersey Autopsy Misses Two Bullets in a Man's Head,"
New York Times, October 20, 1993.

Olson, Carin Met. Cummings, Peter and Rivara, Frederick P. "Association of
First- and Second-Generation Air Bags with Front Occupant Death in Car
Crashes: A Matched Cohort Study," American Journal of Epidemiology 164.2
(2006), 161.

Park, Gene. "FAA Probes Whether Go! Pilots Fell Asleep," Honolulu Star Bulletin
13.51 (2008). Online at http://archives.starbulletin.com/2008/02/20/news/
story08.html. Accessed July 5, 2015.

Parker-Hope, Tara. "A Hollywood Family Takes on Medical Mistakes," New York
Times, March 17, 2008.

Perkins, Lucy. "Human Error Caused Virgin Galactic Crash, Investigators Say,"
National Public Radio, July 28, 2015.

Peters, Jeremy W. "MOTORING; Unraveling the Mystery of Ford's Fire-Prone
Switches," New York Times, August 27, 2006.

Petroski, Henry. To Engineer Is Human: The Role of Failure in Successful Design.
New York: Vantage Books, 1992.

Petroski, Henry. The Evolution of Useful Things. New York: Alfred A. Knopf,
1995.

Petroski, Henry. The Toothpick: Technology and Culture. New York: Alfred
A. Knopf, 2007.

Phillips, Don. "Putting the Pilot Back in Autopilot," The Washington Post
National Weekly Edition, April 29–May 5, 1996.

"Pilot's Wrong Keystroke Led to Crash, Airline Says," New York Times,
August 24, 1996.

Queen v. M'Naghten, 10 Clark & F.200, 2 Eng. Rep. 718 (H.L. 1843).

Report of the Independent Panel of Guidant Corporation, March 20, 2006.

Report of the Presidential Commission on the Space Shuttle Challenger Accident.
Washington, D.C.: United States Government Printing Office, 1986.

Robison, Wade. Decisions in Doubt: The Environment and Public Policy.
Hanover, New Hampshire: University Press of New England, 1994.

Robison, Wade. "In the Moral Zone," Teaching Ethics 8.2 (2008), 57–78.

Robison, Wade. "Design Problems and Ethics," in Goldberg, David E.,
    McCarthy, Natasha and Michelfelder, Diane, eds., *Philosophy and
    Engineering: An Emerging Agenda*, Vol. 2 of the Series Philosophy of
    Engineering and Technology. Springer, 2009.

Robison, Wade. "Ethical Presentations of Data: Tufte and the Morton-Thiokol
    Engineers," in Michelfelder, Diane P., Newberry, Byron and Zhu, Qin, eds.,
    Philosophy and Engineering: Exploring Boundaries, Expanding Connections
    Dordrecht: Springer, forthcoming.

Robison, Wade, et al. "Representation and Misrepresentation: Tufte and the
    Morton-Thikol Engineers on the Challenger," *Science and Engineering Ethics*
    8 (2002), 59–81.

Ryle, Gilbert. *The Concept of Mind*. London: Hutchinson, 1949.

Saul, Roger, Pritz, Howard B., McFadden, Joeseph, Backaitis, Stanley H.,
    Hallenbeck, Heather, and Rhule, Dan. "Description and Performance of the
    Hybrid III Three Year Old, Six-Year- Old and Small Female Test Dummies in
    Restraint System and Out-of- Position Air Bag Environments," Transportation
    Research Center Inc. United States Paper Number 98S7-O–01.

Saul, Stephanie. "Some Sleeping Pill Users Range Far Beyond Bed," *New York
    Times*, March 8, 2008.

Saul, Stephanie. "Ambien in the Driver's Seat," *New York Times*, March 11, 2008.

Saul, Stephanie. "Study Links Ambien Use to Unconscious Food Forays,"
    New York Times, March 14, 2008.

Silberner, Joanne. "Study: Common Heart Drug Combo Raises Risk," *All Things
    Considered*, March 3, 2009.

Sisson, Paul. "Kaiser Hospital Fined for Removing Wrong Kidney," *U-T
    San Diego News*, December 20, 2012.

Stewart, Will. "Russian Roulette Shock as Wedding Guest Shoots Himself in
    Party Trick Gone Wrong," *Daily Mail*, March 23, 2010.

Vaughan, Diane. *The Challenger Launch Decision: Risky Technology, Culture, and
    Deviance at NASA*. Chicago: University of Chicago Press, 1997.

Vlasic, Bill. "G.M. Enquiry Cites Years of Neglect Over Fatal Defect," *New York
    Times*, June 5, 2014.

Voelcker, John. "Who Knew? A Car Battery Is the World's Most Recycled
    Product," *Green Car Reports*, March 31, 2011.

Wald, Matthew L. "Expert Says Engineer Sent Text Messages Before Deadly
    Train Crash," *New York Times*, January 22, 2010.

Wald, Matthew L. and Chang, Kenneth. "Minneapolis Bridge Had Passed
    Inspection," *New York Times*, August 3, 2007.

Wald, Matthew L. and McKinley, Jess. "California Bans Texting by Operators of
    Trains," *New York Times*, September 19, 2008.

Walton, Joseph W. *Essentials of Engineering Design*. St. Paul: West Publishing Co., 1991.

Williams, Timothy. "Citing Failures, Guidant Will Recall Thousands of Defibrillators," *New York Times*, June 17, 2005.

Yaffa, Joshua. "The Road to Clarity," *New York Times Magazine*, August 12, 2007.

INDEX

Note: The letter "n" following locators refers to notes.